W9-ADZ-045

Trace Elements in the Environment

Evaldo L. Kothny, *Editor*

A symposium sponsored by the Division of Water, Air, and Waste Chemistry at the 162nd Meeting of the American Chemical Society, Washington, D.C., Sept. 15, 1971.

ADVANCES IN CHEMISTRY SERIES **123**

AMERICAN CHEMICAL SOCIETY

WASHINGTON, D. C. 1973

ADCSAJ 123 1–149 (1973)

Library of Congress Catalog Card 73–87347

ISBN 8412–0185–4

PRINTED IN THE UNITED STATES OF AMERICA

Advances in Chemistry Series

FOREWORD

Advances in Chemistry Series was founded in 1949 by the American Chemical Society as an outlet for symposia and collections of data in special areas of topical interest that could not be accommodated in the Society's journals. It provides a medium for symposia that would otherwise be fragmented, their papers distributed among several journals or not published at all. Papers are refereed critically according to ACS editorial standards and receive the careful attention and processing characteristic of ACS publications. Papers published in Advances in Chemistry Series are original contributions not published elsewhere in whole or major part and include reports of research as well as reviews since symposia may embrace both types of presentation.

CONTENTS

PREFACE

> We must understand the metabolic process before we condemn the presence of some trace elements with apparently no value.

This volume represents perhaps the first effort to fit geochemistry into environmental science. In studying environmental geochemistry we gain insight into the origin, transition, and concentration of a particular element. Also we may assess the real impact of man-made alterations in natural environments. In fact, the natural cycle of a few elements is altered in smaller environments, but given enough time, they are incorporated back into the whole terrestrial cycle. As the result of this alteration, anomalous metallic concentrations are discovered near cities, freeways, smelters, powerplants, and industrial sites.

Geochemistry is a comparatively young and growing science which studies the chemical composition of the earth. It was pioneered by F. W. Clarke, V. M. Goldschmidt, and V. I. Vernadskii in the late 1920's. Increased growth began with the use of geochemistry as an applied science. Tracing the route and determining the concentration of substances or elements is a most important aspect of applied geochemistry. The first application was in economic geology and in tracing the origin, age, stability, and transformation of certain biogenic and anthropogenic substances.

In the last two decades, anthropogenic emissions of increased amounts of organic substances such as pesticides and other common organic and inorganic pollutants have generated world-wide concern. Emissions enter all three phases: atmosphere, water, and land. They admix and transfer within the phases and during meteorologic phenomena such as duststorms, rain, and oceanic breezes. Tracing the origin of each substance or element becomes increasingly difficult as it ages and is affected by dilution factors related to meteorological, erosional, and tectonic phenomena. For example, in air chemistry the origin of certain emissions is established by studying their concentration, chemical or physical transformation, associated particle size, dilution with air or windblown dust, and the transfer mechanism within the different phases.

Of biological importance are some elements which in low concentrations foster life, such as boron, copper, zinc, molybdenum, cobalt, selenium. At higher concentrations, these elements may be toxic. Therefore, it is important to know their concentration in the natural environment and how these concentrations affect the healthy deployment of life on this planet. Thus, soil chemistry, underground water chemistry, trace element chemistry in plants, and some areas of oceanography and air pollution are interrelated.

Today, most atmospheric toxicants of anthropogenic origin ("pollutants") are organic or generated in the atmosphere by interaction of organic substances with actinic radiation (formation of photochemical smog and ozone). Nevertheless, an important portion of atmospheric toxicants is inorganic, such as sulfur dioxide, carbon oxides, nitrogen oxides, ozone, and sulfuric acid mist. These compounds are the most damaging to plants and humans (chronic bronchiopulmonar diseases). Toxicants in liquid effluents originating from sewage and industrial effluents damage aquatic plants and fish or enter into the food chain of animals and humans. Agricultural sprays and other abatement chemicals add to particulate matter which settles on land plants and may produce chronic or fatal damage to grazing animals and birds. Some substances become concentrated by biological enrichment through the natural food chain. Combustion and manufacturing processes are responsible for emitting large amounts of S, As, V, Zn, Cd, Pb, Cu, Se, Sb, Hg which form part of the metallic content of particulate matter collected in or near cities. Lead is a known contaminant emitted by automotive vehicles. Particle size segregation is a most important tool used to pinpoint the source of these different toxicants.

Geochemical results obtained by studies of mineral exploration, meteorology, oceanography, biochemistry, agriculture, air pollution, and others have been scattered through the meetings of several different societies. The Association of Exploration Geochemists was founded in 1970 for fostering symposia on that specific subject. The science of trace elements in agriculture is comparatively young and expands constantly as a spinoff of fertilizer chemistry. Most information about trace elements in agricultural soils and their availability to the plants has been obtained in East European countries. In the classical sciences of soil and water chemistry, the study of trace constituents is relatively new. Outstanding work has been accomplished in this area by the U.S. Geological Survey under title of "The Geochemical Series." Two large groups are directly interested in soil and water chemistry. The first is concerned about agricultural production and the second about economic geology, such as petroleum, mineral, and geothermal exploration.

Because environmental geochemistry is implicitly but not exclusively a part of the Division of Air, Water and Waste Chemistry (now changed to Division of Environmental Chemistry), a symposium on that subject was assembled as a part of this Division under the title "Geochemical Cycle of Trace Elements in our Environment." However, this was not the first symposium on geochemistry organized by the American Chemical Society. A symposium "On Problems in Analytical Geochemistry" was presented at the SE-SW Regional Meeting in Louisiana, December 1970. As a result of the papers presented at the 162nd National Meeting of the American Chemical Society in Washington, D.C. in September 1971, this book was compiled.

We sincerely hope that the scattered environmental geochemical information may be combined in the future to broaden the knowledge about the routes of transfer of each constituent, either from natural sources or emitted by human activity and the ultimate effect on biological systems. In publishing these results, a start in this direction has been made which may eventually lead to a fertilization in this line of thought.

Berkeley, Calif.
July 1973

EVALDO L. KOTHNY

1

Lead Source Identification by Multi-Element Analysis of Diurnal Samples of Ambient Air

J. J. WESOLOWSKI,[a] W. JOHN,[b] and R. KAIFER

University of California, Lawrence Livermore Laboratory,
Livermore, Calif. 94550

A method for determining the existence and nature of non-automotive lead sources in a given area from ambient air particulate samples has been developed. The method consists of first measuring the diurnal variations of the Br/Pb ratio using x-ray fluorescence to determine those days on which the non-automotive lead source is operating. For the episodal days the diurnal concentration patterns of about 20 metals are measured using neutron activation analysis. The correlation of these patterns with the lead pattern characterizes the source. The method has been tested in Benicia, Calif. and has successfully established the existence of a non-automotive lead source in that area and characterized the type of source.

Measurements of lead (Pb) concentrations in 24-hour high volume filters obtained from monitoring stations are of little value in distinguishing between automotive and non-automotive Pb sources. Mainly, the time averaging is too long to be effective in spotting short-term source emissions, especially since these will be superimposed on a large and relatively constant automotive Pb background. Yet determining the existence and the nature of non-automotive Pb sources in a given area can be crucial under certain circumstances. Such circumstances were realized in Solano County, Calif. in 1970 when investigations by the California State Department of Public Health led to the conclusion that a number of horses which had died in a certain area of the county had been suffering from lead poisoning at the time of their deaths (*1*). Sub-

[a] Present address: California State Department of Public Health, 2151 Berkeley Way, Berkeley, Calif. 94704.
[b] Present address: Department of Physical Sciences, California State College, Stanislaus, Turlock, Calif. 95380.

sequent investigations by a number of state agencies established a Pb pollution problem in the vegetation of the area and determined that this Pb is primarily deposited from the air (*1*). The polluted area in question is semirural and located approximately 20 miles northeast of San Francisco near the city of Benicia. It is part of the Carquinez Straits which form a natural corridor for the venting of airborne contaminents from the San Francisco Bay Area and is also central to a large industrial complex. Thus, it was imperative to distinguish between automotive and non-automotive Pb in ambient air. This paper is concerned only with the technique developed to determine the existence and nature of a non-automotive Pb source in a given area. A comprehensive review of the Benicia pollution problem can be found in Ref. *1*.

Sampling Technique

As obtained by various state agencies using 24-hour high volume measurements, Pb concentrations in Benicia from March to May 1970, ranged between 0.024 and 1.3 μgram/m^3—a range which is lower than that found for a similar period the previous year in the nearby urban area of San Francisco (*2*). Thus, the high Pb concentrations in the vegetation must be caused by circumstances peculiar to the area. These could include source episodes not coincident with measurement periods, different particle size distributions than usually encountered in an urban area, and higher wind velocities than obtained in an urban area causing greater impaction efficiency. Further, the extremely complicated and time varying meteorology of the Bay Area made any source direction judgments based on 24-hour averaging very hazardous. Clearly the measurement of Pb concentration variations as a function of time for small time intervals would establish episodal patterns and, when correlated with wind data, would help determine source location. Unfortunately, such measurements by themselves do not distinguish between automotive and non-automotive sources. Since the primary source of bromine (Br) in the area is the automobile, we believed that the measurement of the Br-to-Pb ratio would make possible this distinction. [The Br contribution from ocean aerosol, for the purposes of this paper, can be neglected. A rough measure of the effect was obtained by assuming a Na/Br seawater value and calculating the ocean's Br contribution from the measured concentration of sodium.] If the existence of a non-automotive source were established, then the signature of the specific polluter could be obtained by a multi-element analysis of the samples.

Thus in November 1970 we began to monitor the air in the Benicia area in two-hour time intervals for three days each week with the anticipation that the analysis of the concentration variations of many elements might answer three questions:

(1) Is some of the Pb in the air of non-automotive origin?

(2) In what direction with respect to the sampler are the sources located?

(3) What is the nature of these sources?

The sampling was carried out with an automatic sequential sampler designed and fabricated at LLL. Twelve samples, each of two-hours duration could be collected sequentially before the unit needed servicing.

Each sample was taken on a 29-mm diameter Whatman No. 41 filter paper using a 1/3 hp vacuum pump operated at a constant flow rate of 1.5 cfm (43 lpm). This provided a sample size of 5.1 m^3 collected at a linear velocity of 143 cm/sec, at which speed aerosol collection efficiencies exceed 98% down to diameters smaller than 0.1 μm (*3*).

The sequential sampler was positioned about 25 feet above ground, overlooking a small private yacht harbor at Elliot Cove, Calif. from November 13, 1970 to March 1, 1971 (Designated as station M-4 in Ref. *1*).

The samples automatically sequenced three times a week, 1600 hours Monday to 1600 hours Tuesday, 1600 hours Wednesday to 1600 hours Thursday, and 1600 hours Friday to 1600 hours Saturday. Samples were collected with the cooperation of the Air Resources Board and the Air and Industrial Hygiene Laboratory of the Department of Public Health.

Analytical Procedure

The Pb and Br diurnal patterns were measured for a particular day using x-ray fluorescence techniques (*4*). Those patterns which demonstrated a possible non-automotive Pb component were then analyzed for many elements including Br but not Pb, by neutron activation analysis (NAA) (*5*). Measurements of Br concentrations by the two techniques were within experimental error. Both techniques are non-destructive.

The x-ray fluorescence system consisted of a radioactive source as the excitor and a 200-ev resolution (FWHM) lithium-drifted silicon detector with standard electronics. Only 15 minutes were required to completely process a sample.

The neutron activation analysis was carried out at Livermore's 3-megawatt reactor. High resolution lithium-drifted germanium detectors were used to count the samples. Details of the irradiation and counting sequence can be found in an earlier paper (*6*).

Results

The Br/Pb ratio can help distinguish between automotive and non-automotive Pb. Anti-knock additives usually contain ethylene dibromide which reacts with Pb deposits in the engine to form Pb halide particles which are emitted in the exhaust. The initial Br/Pb weight ratio is 0.39. Unfortunately, because of the labile nature of Br, the automotive particulate Br/Pb ratio is not constant but decreases with time after emission (7). To approximate the range of the ratio, Pb measurements were made on two-hour samples collected over three typical summer days in Livermore, Calif. in July 1970. Earlier, measurements for elements using NAA (including Br but not Pb) had demonstrated that a substantial amount of the Livermore smog is "imported"—*i.e.*, is an aged aerosol (*6*). This coupled with the lack of known non-automotive lead sources in the area makes Livermore a good location for studying the range of the automotive particulate Br/Pb ratio. Figure 1 shows the results of these measurements. Error bars refer to relative analytical errors only. The correlation between the concentrations of Pb and Br is excellent, indicating a common source. Note also that the Br/Pb ratio varies, for the

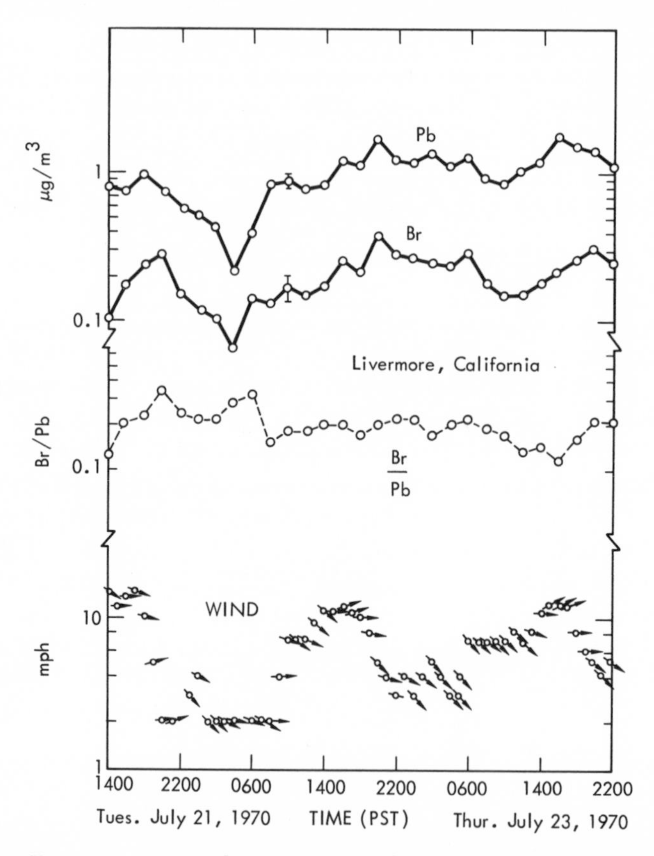

Figure 1. Diurnal variations of Pb, Br, and wind for July 21, 22, 23, 1970 in Livermore

most part, over a small range of values and is always greater than 0.1. The absolute concentration of Pb is never greater than 2 μgrams/m³, and the average value for the two days is 1.1 μgram/m³. The arrow depicting wind direction points in the direction the wind is blowing. The top of Figure 1 is north.

Most of the Benicia samples analyzed gave Br/Pb ratios which indicated that the major component of the Pb aerosol was automotive in origin. Figures 2 and 3 give some representative diurnal patterns.

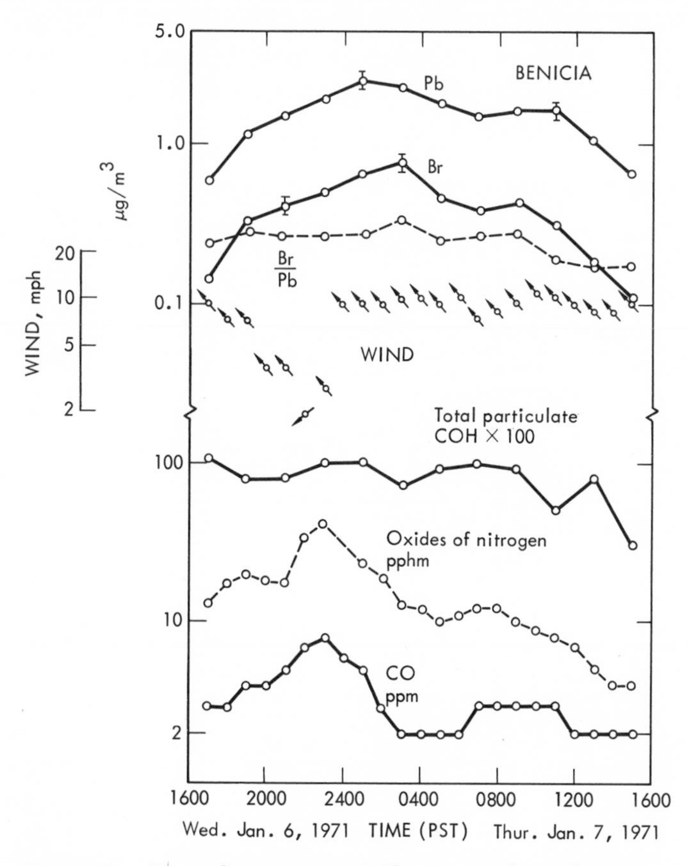

Figure 2. Diurnal variations of Pb, Br, wind, total particulate, and gases for Jan. 6, 7, 1971

Figure 2 is for January 6-7, 1971, when the winds were mostly from the southeast. Wind data were obtained from a station near the Carquinez Bridge located about two miles west of the sampler. Particulate and gaseous data were obtained from the State of California Air Resources Board station in the city of Vallejo, approximately five miles northwest of the sampler. The correlation between Pb and Br is good, and the Br/Pb ratio varies from 0.17 to 0.34. The average value of the Pb concentration is 1.51 μgrams/m^3.

Figure 3 is for February 3-4, 1971. The wind varied from SW to NE (data is from a station five miles northeast of the sampler since none was available from the bridge). The Br/Pb ratio varies between 0.1 and 0.3. In fact there seems to be a general correlation between total particulates, CO, Br, and Pb. Oxides-of-nitrogen data were not available. The average Pb concentration is 2.02 μgrams/m^3.

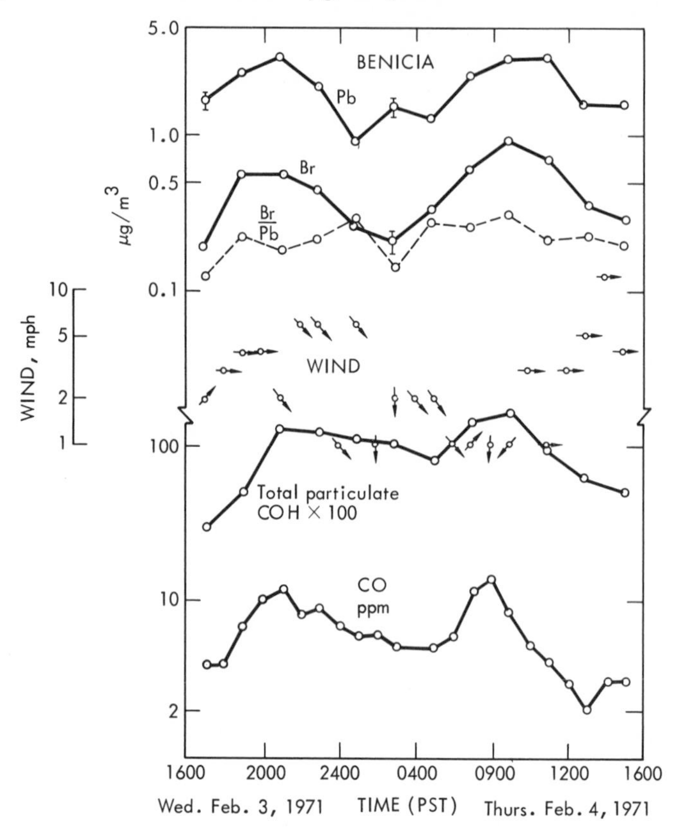

Figure 3. Diurnal variations of Pb, Br, wind, total particulate, and CO for Feb. 3, 4, 1971

For these days and many others not shown, the results are sufficiently similar in character to those from Livermore that one can conclude that an automotive source alone would be sufficient to explain the Pb concentrations.

To substantiate the above conclusion, samples from a number of suspected predominately automotive source days were multi-element

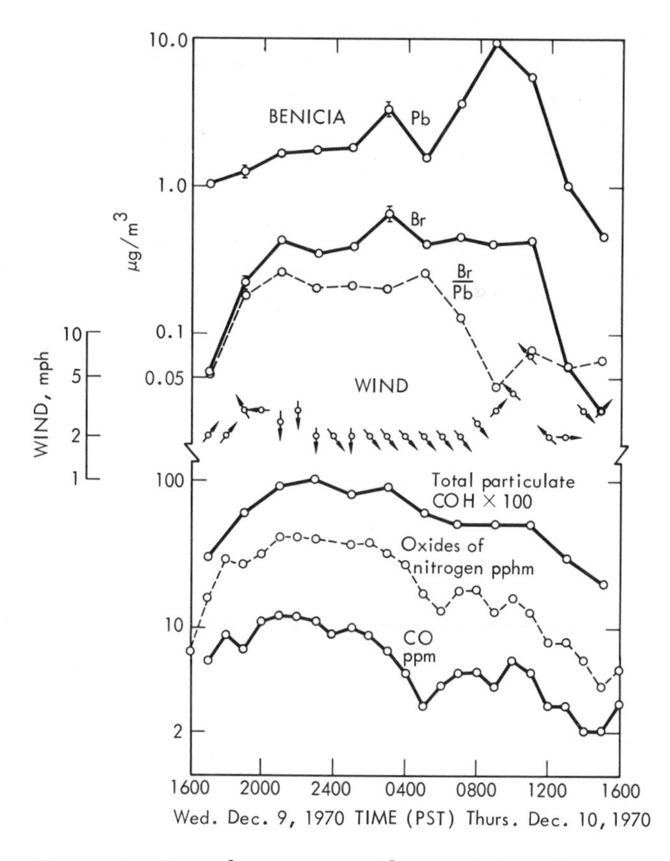

Figure 4. Diurnal variations of Pb, Br, wind, total particulate, and gases for Dec. 9, 10, 1970

analyzed using NAA. In spite of the light loading of the two-hour samples, 28 elements were detected. If the major Pb component were caused by a non-automotive source such as a smelter, one would expect trace metals other than Pb to be present in concentrations larger than normally found in the area. In general, this was not the case. For example, on January 6-7 (Figure 2) the maximum concentration for arsenic (As) during the 24-hour period was 0.005 μgram/m^3, and for antimony (Sb), 0.004 μgram/m^3. The results for these and most other metals are consistent with those obtained from the neutron activation analysis of 29 elements from 24-hour high volume filters from nine Bay Area Air Pollution Control District stations for a typical summer day in July of 1970 (*8*). The above results do not imply that no non-automotive Pb source was operating but that any such sources contributed only a minor component to the Pb aerosol concentrations.

However, there were three days for which the diurnal Br/Pb patterns were different from those above. The diurnal patterns for these three days have at least three things in common.

(1) Even though their 24-hour average Pb concentrations are not excessively high, they have large Pb concentration excursions for part of the day.

(2) The Pb excursion is not accompanied by a corresponding increase in Br concentration—*i.e.*, the Br/Pb ratio drops during the excursion.

(3) All three excursions were preceded by winds from the N to NW sector.

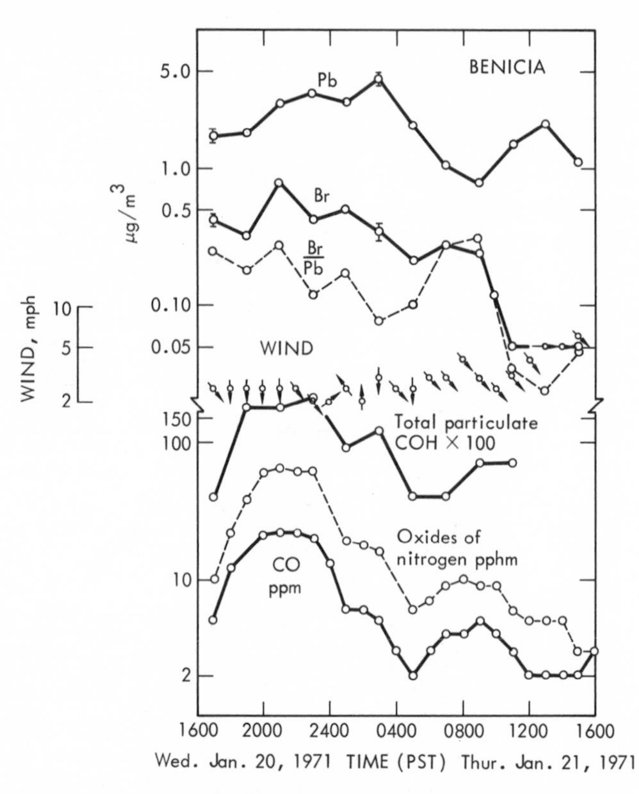

Figure 5. Diurnal variations of Pb, Br, wind, total particulate, and gases for Jan. 20, 21, 1971

The most dramatic pattern is shown by the data of December 9-10, 1970, Figure 4. Although the Br diurnal pattern approximates that for total particles which is consistent with an extended Br source—*viz.*, the automobile—the Pb pattern shows in addition a large peak on the morning of December 10. At 0900 of this day the average Pb concentration for the two-hour period was 9.4 μgrams/m³, yet the Br/Pb ratio dropped to a low value of 0.042. When the wind changed from NW to SE and increased in velocity, the concentration levels in general drop sharply.

The pattern of January 20-21, 1971 (Figure 5), shows a peak Pb concentration of 4.5 μgrams/m^3 with a corresponding Br/Pb ratio of 0.076. The Br/Pb ratio increases to the typical automotive value of 0.3 as the Pb concentration drops down to the more "normal" value of 1 μgram/m^3. Also on this day a second Pb excursion reached a value of 2.1 μgrams/m^3 with a Br/Pb ratio of 0.024.

November 16-17, 1970 (Figure 6) has a high Pb value of 4.2 μgrams/m^3 at 2100 November 16, with a corresponding Br/Pb ratio of only 0.039. There are two more Pb peaks at 0100 and 0900 on November 17. In both cases these peaks are accompanied by decreases in the Br/Pb ratio.

The x-ray fluorescence analysis thus selected these three days as candidates for days on which a major component of the Pb aerosol was from a non-automotive source. To establish that this was the case and actually to determine the nature of the non-automotive source, a multi-

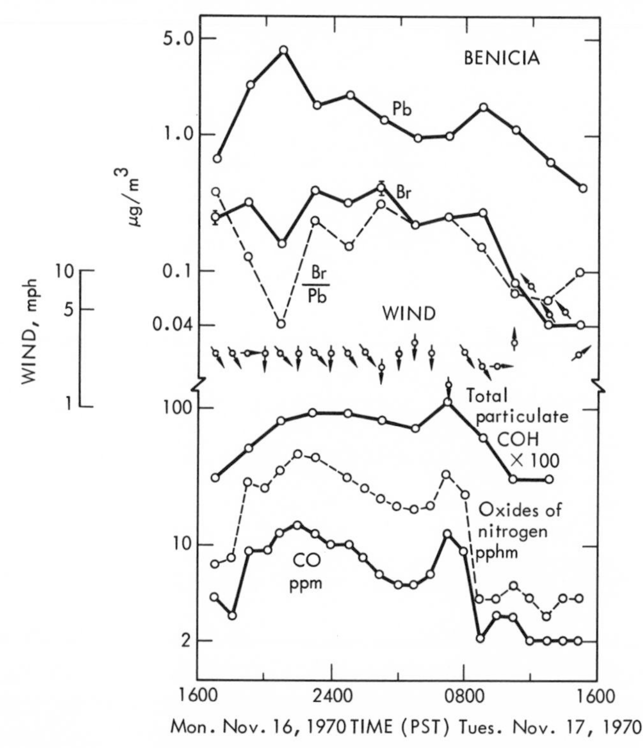

Figure 6. Diurnal variations of Pb, Br, wind, total particulate, and gases for Nov. 16, 17, 1970

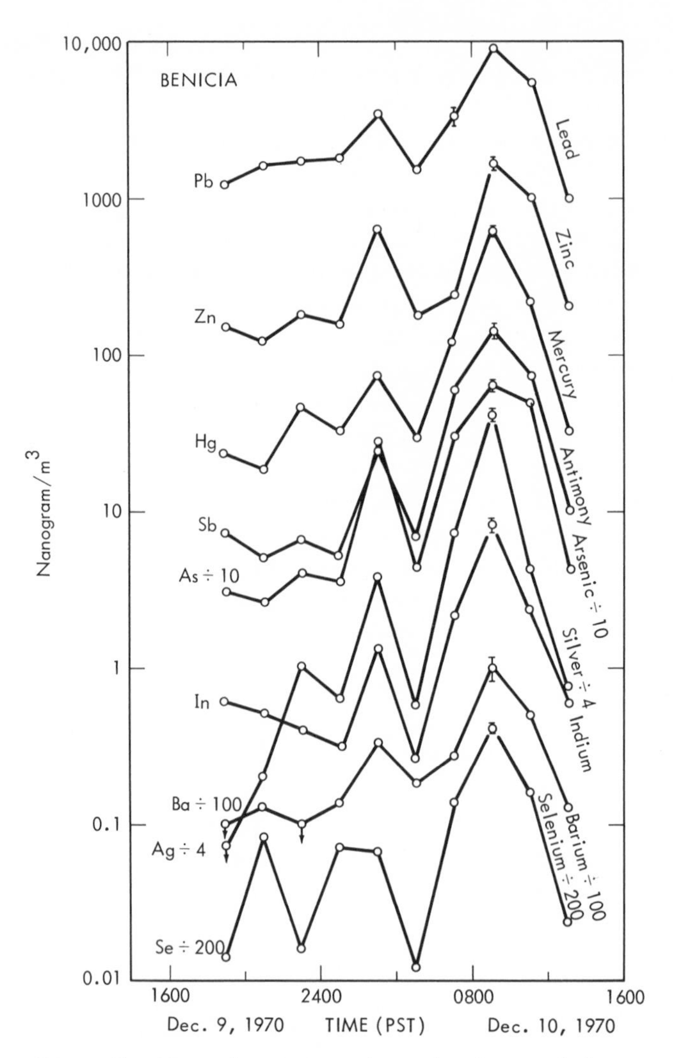

Figure 7. Diurnal variations of correlated aerosol elemental concentrations for Dec. 9, 10, 1970. The symbol with the arrow indicates the element was not detected. The symbol is placed at the lower limit of detection.

element analysis on all the samples for these three days was carried out using NAA.

For December 9–10, 26 elements were detected: Na, Al, Cl, K, Ca, Sc, V, Cr, Mn, Fe, Co, Cu, Zn, As, Se, Br, Ag, Cd, In, Sb, I, Ba, La, Sm, W, and Hg. Those elements whose diurnal patterns were strongly correlated to lead are shown in Figure 7. The absolute concentrations of these elements at the peak of the episode are also much larger than those normally encountered in this area. Table I lists these peak concentrations for the correlated elements. Those elements with little or no correlation are shown in Figure 8. Sensitivities were not sufficient to obtain complete diurnal patterns for K, Ca, Sc, Cr, Fe, Co, Cu, Cd, I, La, Sm, and W. However, except for Cu, Cd, and W, the patterns were complete enough to determine that none of these elements correlated with Pb. Further, their absolute concentrations were normal for the area. Cadmium (Cd) was detected on only three filters. A maximum value of 120 ngrams/m^3 was detected at 1100 on December 10. The detection limit for Cd of about 30 ngrams/m^3 depends to some extent on the amount of other metals present. Thus no definite conclusion can be drawn regarding the possible correlation of Pb and Cd for this day.

It is interesting that the Cl and Na patterns are not well correlated and that for most of the 24 hours the Cl/Na ratio is much greater than the seawater value of 1.8. In the Livermore study (*6*) the patterns were well correlated, and the variation of the Cl/Na ratio was a factor of two with a maximum value of 1.6. The anomalous Na and Cl behavior for December 9-10 was repeated in Benicia on other days. This indicates the possible presence of an anthropogenic source of chlorine (Cl) in the area.

For the January 20-21 episode those elements which correlated with Pb are shown in Figure 9. Again the concentrations at the episode peak are large (*cf.*, Table I). Many other elements were detected but showed no strong correlation with Pb. There was a weak correlation between Pb and Na, but the Na concentrations were normal for the area. No Cd was detected. Barium (Ba) was detected on five filters with a concentration range of 15 to 35 ngrams/m^3—the largest value being observed at the episode peak. Selenium (Se) was detected on five filters with values ranging from 2.1 to 9.2 ngrams/m^3. The maximum value did not occur at the episode peak.

Table I. Concentrations at Episode Peak for Correlated Elements, nanograms/m^3

Element	*2100 Nov. 16*	*0900 Dec. 10*	*0300 Jan. 21*
Pb	4150	9400	4500
Zn	1160	1770	1840
Hg	44	630	45
Ba	33	104	35
Se	11	82	5
As	176	630	1400
Sb	97	148	760
In	3.1	8.6	1.9
Ag	4.5	175	58
Cd	290	<30	<30

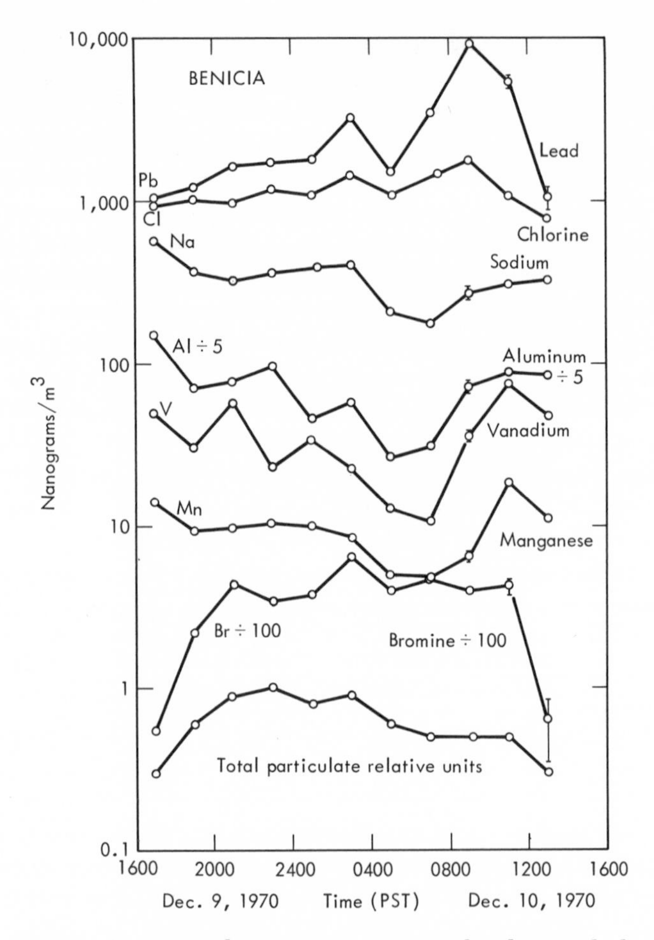

Figure 8. Diurnal variations of uncorrelated aerosol elemental concentrations for Dec. 9, 10, 1970

The diurnal patterns for Zn, Sb, As, and In are almost identical. They correlate reasonably well with Pb but do have a more resonant behavior. This is particularly noticeable on the morning of January 21 when their concentrations are about a factor of 20 lower than they were at the episode peak. It is most interesting that an "automotive background" subtraction of around 0.8 μgram/m^3 from the Pb data would cause the Pb diurnal pattern to be very similar to the Zn, Sb, As, In patterns.

Figure 10 shows the diurnal patterns for those elements correlated with Pb on November 16-17, 1970. Sensitivities were not sufficient to obtain complete patterns for Cd and Hg. Diurnal patterns for Ag, Se, and Ba were incomplete. The maximum values observed for Ag, Se, and Ba were 4.5, 10, and 35 ngrams/m^3. The concentration of As at the peak is 1/10th that for January 20-21 (*cf.* Table I). Further, the diurnal patterns for most of these elements are even more resonant than those for December 9-10 or January 20-21. For example, the Zn concentration varies by a factor of 400 between its maximum and minimum values. Further, the minimum values achieved during the morning hours of November 17 would be considered typical for the area. For example, the minimum Sb value of 1.4 ngrams/m^3 is in good agreement with the average value of 2.0 ngrams/m^3 for the nine station Bay Area Survey of 1969 (*8*). As on January 20-21, an automotive background would bring the Pb diurnal pattern in closer agreement with the other elements.

The above data not only confirm the existence of a non-automotive Pb source but also determine the nature of the source. The correlated elements for the three episodes described would be consistent with a smelter source. Indeed a smelter is located about two miles WSW from the sampler. However, correlation of the ground level wind data with

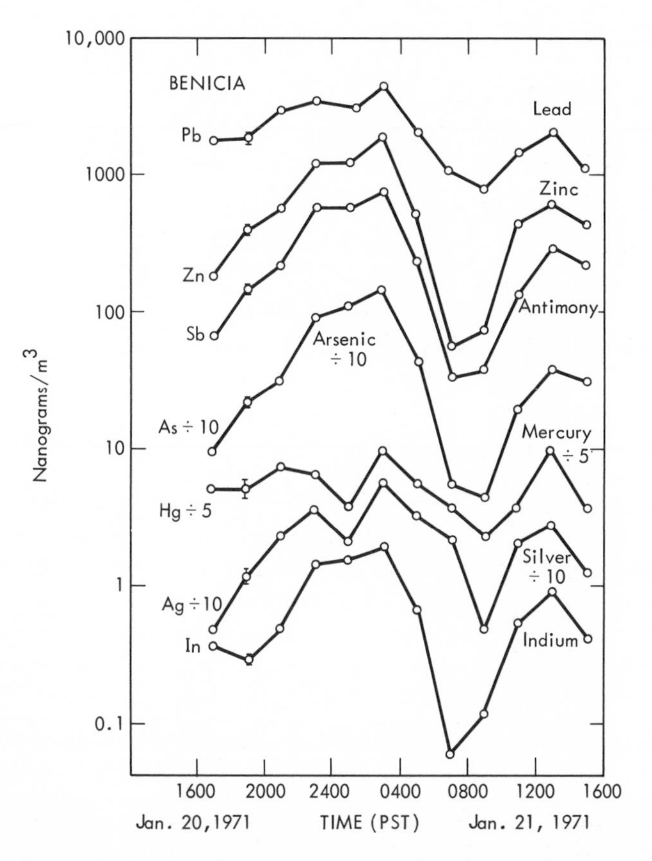

Figure 9. Diurnal variations of correlated aerosol elemental concentrations for Jan. 20, 21, 1971

the elemental diurnal patterns indicates a NNW direction. This discrepancy is not surprising for two reasons. First, the stacks of the smelter in question are 100 and 600 feet high. Secondly, the topography and meteorology of the San Francisco Bay Area are so complicated as to

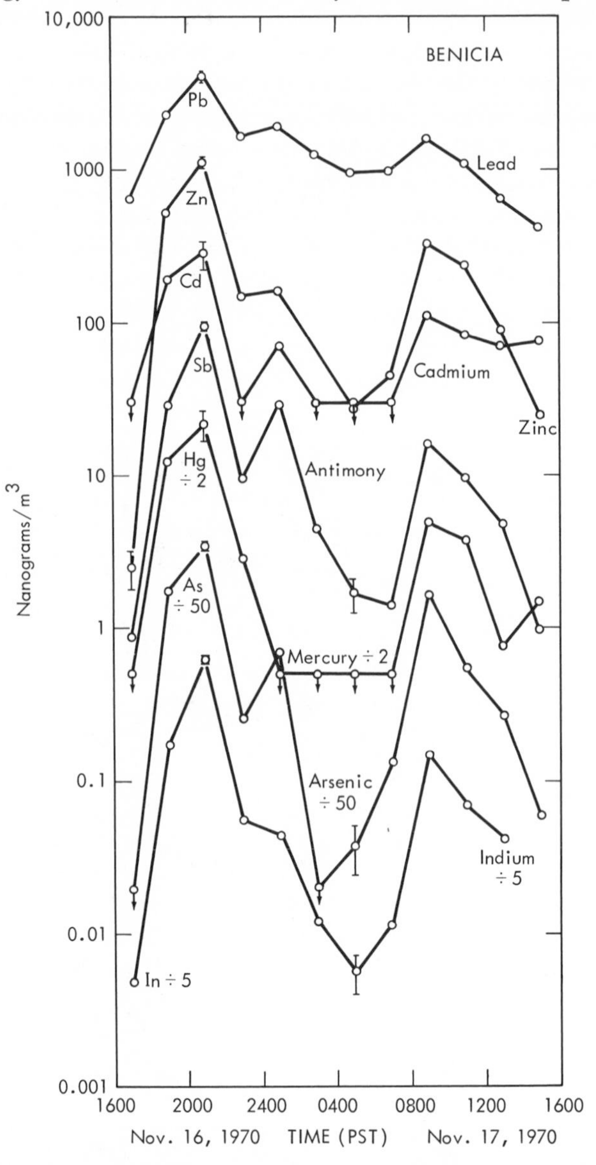

Figure 10. Diurnal variations of correlated aerosol elemental concentrations for Nov. 16, 17, 1970. The symbol with the arrow indicates the element was not detected. The symbol is placed at the lower limit of detection.

make flow pattern predictions based on ground level measurements at one station very uncertain.

Attempts to obtain daily production records of the smelter, to correlate emissions with the diurnal patterns shown above, were unsuccessful. However, indications were that the smelter was in the process of "shutting down" during this experiment. This, coupled with the varying meteorological conditions, could be the reason more episodal days were not found during the experiment. The smelter processed ores from various locations and extracted such metals as Pb, Zn, Au, Ag, and Pt (*9*). At the present writing the smelter is not operating.

In April of 1970 a source test of the 600-foot stack was carried out by the Bay Area Air Pollution Control District (*1*). The results which were given in lbs/day are presented in Table II normalized to As to make meaningful comparisons with the diurnal data of this experiment. The qualitative agreement between stack and ambient air data further implicates the smelter as the source of the episodes on these three days.

Table II. Comparison of Typical Stack Sample Concentrations with Ambient Air Sample Concentrations at Episode Peak Normalized to Arsenic

Element Ratio	*Stack*	*Nov. 16-17*	*Dec. 9-10*	*Jan. 20-21*
As/Zn	0.31	0.15	0.36	0.76
As/Pb	0.17	0.042	0.067	0.31
As/Cd	0.40	0.61	—	—

Summary

A method for determining the existence and nature of non-automotive Pb sources in a given area from ambient air particulate samples has been developed. The method consists of first measuring the diurnal variations of the Br/Pb ratio using x-ray fluorescence to determine those days on which the non-automotive Pb source is operating. For the episodal days the diurnal concentration patterns of about 20 metals are measured using neutron activation analysis. The correlation of these patterns with the Pb pattern characterizes the source. Further evidence regarding the identification of the source can be obtained by comparing the ratios of the concentrations of the correlated elements with stack sample data. The method has been tested in Benicia, Calif., and has successfully established the existence of a non-automotive Pb source in that area and characterized the type of source. The location of the source with respect to the sampling station can in principle be inferred by correlating wind direction with the elemental diurnal patterns. To be useful, the wind data must be more comprehensive than that available for this experiment.

Although this paper has stressed a lead pollution problem, it should be clear that the comparison of the diurnal concentration variations of many elements is a powerful method in general for determining ambient air pollution problems in a given area.

Acknowledgments

We thank P. K. Mueller and the staff of the Air and Industrial Hygiene Laboratory of the Department of Public Health for assistance in changing our samples; K. Jones of Air Resources Board for his help in setting up the sampling station; J. Koslow and the staff of ARB for gas and wind data; M. Feldstein and W. Siu of Bay Area Air Pollution Control District for wind data. We also thank J. Tinney, S. Chin and J. Cate, LLL, for the use of counting equipment; H. Palmquist, W. Wade and H. Chesnutt, LLL, for the design and servicing of the sampler. We wish to express our appreciation to C. Maninger for encouragement and support.

Literature Cited

1. "A Joint Study of Pb Contamination Relative to Horse Deaths in the Area of Southern Solano County," State of Calif. Air Resources Board Report, Dec. 1971.
2. Mueller, Peter K., Stanley, Ronald L., "Origin of Lead in Surface Vegetation," State of California Dept. of Public Health, Air and Industrial Hygiene Laboratory Report No. **87** (1970).
3. Lindeken, C. L., Morgin, R. L., Petrock, K. F., "Collection Efficiency of Whatman 41 Filter Paper for Submicron Aerosols," *Health Phys.* (1963) **9,** 305.
4. Cate, Jr., J. L., "Determination of Lead in Air Sample Filters by X-Ray Fluorescence Analysis," *Univer. Calif., Lawrence Rad. Lab. Rept.* **UCRL-51038.**
5. Gordon, G. E., "Instrumental Activation Analysis of Atmospheric Pollutants and Pollution Source Materials," International Symposium on Identification and Measurement of Environmental Pollutants, Ottawa, Ontario, Canada, 1971.
6. Rahn, K., Wesolowski, J. J., John, W., Ralston, H. R., "Diurnal Variation of Aerosol Trace Element Concentrations In Livermore, California," *J. Air Pollut. Control Assoc.* (1971) **21,** 406.
7. Winchester, John W., Zoller, William H., Duce, Robert A., Benson, Carl S., "Lead and Halogens in Pollution Aerosols and Snow from Fairbanks, Alaska," *Atmos. Environ.* (1967) **1,** 105–119.
8. John, W., Kaifer, R., Rahn, K., Wesolowski, J. J., "Trace Element Concentrations in Aerosols from the San Francisco Bay Area," *Atmos. Environ.* (1973) **7,** 107–118.
9. "Air Pollution in the San Francisco Bay Area," Stanford Workshop on Air Pollution, Ecology Center Press, San Francisco, 1970.

RECEIVED January 7, 1972. Work performed under the auspices of the U. S. Atomic Energy Commission.

2

Geochemical Aspects of Inorganic Aerosols near the Ocean–Atmosphere Interface

D. J. BRESSAN, R. A. CARR, and P. E. WILKNISS

Naval Research Laboratory, Washington, D. C. 20390

Sources contributing to the composition of inorganic aerosols near the ocean–atmosphere interface are the oceans themselves, continental dust, volcanic ash, atmospheric production of particulates, and, to lesser extents, human activity and extraterrestrial inputs. Characteristic elements and elemental ratios can be used to determine some of these sources and detect ion fractionation at the sea–air interface. Rain water chemistry is not always simply related to that of the marine aerosol.

The chemical composition of inorganic aerosols near the sea–air interface varies widely with geographical location and existing meteorological conditions (Figure 1). The main component of these aerosols up to an altitude of approximately 2 km is "sea salt," including marine surfactants, derived from the ocean surface (*1–4*). Most of the sea salt component is directly returned to the oceans, either by fallout or precipitation. Part of it is transported to the continents where its fate has not been studied in detail (*5*). There is also a continental component in the over-ocean aerosols, sometimes significant, which consists of airborne dust transported from the continents out over the oceans by prevailing winds. This dust transport is now a recognized mechanism for the addition of trace metals to oceanic sediments and in certain cases can be of considerable importance in overall sediment accumulation rates (*6*). In areas of volcanic activity, contributions to the aerosol may be great during periods of active eruptions. Heavy volcanic contributions are short-lived events, at least near the sea–air interface, but are well documented by ash layers found in deep-ocean sediments (*7, 8*). The atmosphere itself has been recognized as a source of aerosols, namely when gases such as H_2S or SO_2 are oxidized to form particulates (*9*). The contribution of extraterrestrial material to the atmospheric aerosols

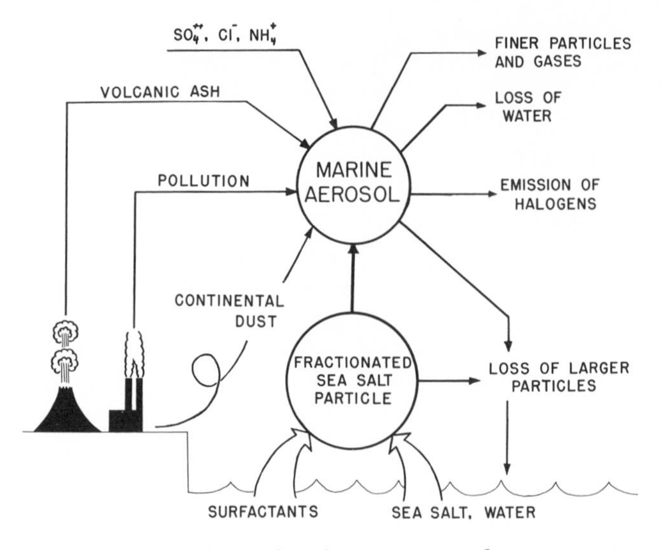

Figure 1. A marine aerosol

is not well investigated and is of lesser significance (*10*). Human activities are also a contributing source to marine aerosols as demonstrated by investigations concerning lead aerosols in the marine atmosphere (*11*).

The geochemical aspects of inorganic aerosols in the marine atmosphere pose a formidable problem for scientific investigations. So far, we do not have one complete chemical analysis of such an aerosol, that is, one comprising all cations and anions. According to their interests, investigators have concentrated either on anions such as Cl^-, Br^-, I^-, SO_4^{2-}, PO_4^{3-}, and NO_3^- or on cations such as NH_4^+, Na^+, K^+, and Ca^{2+}. The available data for the transition metals are too scattered and fragmentary to allow detailed geochemical conclusions.

Sampling and Analysis of Marine Aerosols

Marine aerosols are sampled for dust and salt by impingement of the aerosol particles onto nylon meshes (*12*). This "dust-kite" method (*13, 14*) uses a mesh with a strand diameter of 230μ and an air speed of approximately 15 knots. High-speed aerosol collectors (*12*) used on ships and aircraft employ meshes with strands of 48μ diameter and an air speed of 75–150 knots. Electron micrographs show that dust particles of $\geq 0.1\mu$ are collected by these high-speed samplers. To collect condensates (rain and fog) aboard ship, we have modified the shape of a dust kite so that the condensates will run to one point on the bottom and drop directly into a 1-liter polyethylene bottle. The system has collected up to 500 ml/hour in a fog of 0.5-mile visibility.

Other methods of collecting atmospheric aerosols include the scrubbing tower (*15, 16*), which will also trap gases and various types of cascade impactors, used to differentiate particles by size (*15–18*). Different widths of wire ribbon can be used to collect aerosols by impingement and to differentiate particle sizes. They must have high air-flow rates and have been used in conjunction with aircraft (*15*). Filter papers have been used with high-volume air samplers to collect atmospheric aerosols containing both inorganic and organic matter (*19, 20*). Treated papers can be used for the specific analysis of lead (*21*). Cloud chamber devices have been used in cloud physics work to sample, count, and indirectly analyze airborne particles which act as condensation nuclei (*22, 23*).

Rainwater samples have been used as an indirect method of aerosol sampling (*24*), but the presence of gaseous atmospheric chlorine (*25, 26*) could invalidate to some degree the estimations made from such samples of Cl^- in the aerosols (*25, 27*). Some aerosol particles can be selectively removed by nucleation, condensation, and precipitation scavanging, while gaseous components can be absorbed by the rain drops. The elemental ratios in the rain would then be different from those in the solid atmospheric aerosol (*24*).

In the analysis of a marine aerosol, the first step is to divide it into a water-soluble fraction which we call salt and an insoluble fraction referred to as dust. This is achieved by washing the collecting nylon "sail" or filter with distilled water (*12*). The dust is removed from the suspension by centrifugation and filtration. It is dried at 110°C and weighed. The salt is either isolated from an aliquot of the supernate by freeze drying, or the dissolved ions of interest are determined directly. The amount of salt in each sample is determined by weighing or by summing the major constituents.

The analyses of the dust and salt components have been performed by various methods in our laboratory. Neutron activation analysis was used to determine Na, Mg, Al, Ti, V, Mn, Sc, Fe, Co, Sb, Ce, and Eu in dust (*28*). Fluoride in dust (and the K value in Table III) was done by photon activation analysis (*12, 28*). The salt portions of the aerosol and rainwater samples were analyzed for Na, K, Cl, and F. Sodium was determined by neutron activation and atomic absorption, potassium by atomic absorption, chlorine by titration, photon activation, and neutron activation, and fluoride by ion-specific electrode and photon activation (*12, 28–30*). Two analytical methods were used for Na, F, and Cl in the same samples to determine precision and accuracy.

I and Br have also been analyzed in aerosols and rain by neutron activation (*15*). The major chemical components may also be estimated from the mineralogical composition indicated by x-ray diffraction techniques (*13*).

Table I. Marine

Area Sampled	Wind Speed, knots	Wind Direction, degree
Atlantic, 15°N, 53°W [a]	14–19	070
Atlantic, 27°N, 30°W [a]	15–20	070
Pacific, 33°N, 165°W [a]	5–15	120
Pacific, 27°N, 161°W [a]	5–12	100
Caribbean, 16°N, 70°W [a]	12–16	090
Caribbean, 15°N, 68°W [a]	15–20	075
Greenland Sea, [a] 75°N, 01°E	3–8	320
Greenland Sea, [a] 75°N, 06°E	8–14	120
East Coast, 39°N, 69°W [a]	10	190
Atlantic, 18°N, 50°W [a]	11–14	050
Atlantic, 25°N, 35°W [a]	12–14	080
Atlantic, 26°N, 33°W [a]	9–14	070
Antartic, 48°S, 170°E [b]	20–30	315
Antarctic, 50°S, 170°E [b]	15–25	315
Antarctic, 55°S, 170°E [b]	10–15	315
Antarctic, 60°S, 170°E [b]	15–30	000
Antarctic, 65°S, 170°E [b]	20–25	000

[a] Authors' data.

Composition of Marine Aerosols with Respect to Geographic Area

The compositions of marine aerosols with respect to salt and dust are shown in Table I. The data show the presence of dust in every sample. The dust component ranges from a fraction of 1% to about 30% by weight. Aerosol particles with diameters greater than 0.2μ contain the major mass of a marine aerosol (*31*). Apparently the dust contribution does not depend on the weather at the time of sampling. Rather, it seems to be related to the location and the general movement of air masses. This is best shown for the equatorial Atlantic where the heaviest dust loadings are found. It has been demonstrated by others (*32*) that this dust originates in Africa and is carried out over the ocean by the prevailing winds.

The concentration of salt in the marine aerosols depends strongly on the local wind force during sampling as shown in Table I and as documented by other investigators (*31, 33*). The salt is locally derived at the sea surface from white caps and spray associated with wind action.

Included with the salt is marine surfactant material (*4, 19, 34*). The concentration of surfactant at the sea suface appears greatest in sea slicks (*35*), and its composition and degree of surface activity depend on various biological factors (*36*). Transition metals have been shown to be enriched in these materials (*4, 37*) even when more than 80% of the

Aerosol Composition

Particle Diameter, μmeters	*"Salt," μgram/meter³*	*"Dust," μgram/meter³*	*Salt/Dust*
≧2	15.4	0.02	770
≧2	15.0	3.8	4
≧2	1.0	0.08	13
≧2	0.6	0.02	30
≧2	6.5	0.16	40
≧2	10.7	0.42	25
≧2	0.9	0.005	180
≧2	3.5	0.1	35
≧0.2	12.0	1.3	9
≧0.2	8.6	1.6	5
≧0.2	28.0	6.4	4
≧0.2	12.0	3.8	3
—	0.78	.012	65
—	1.95	.013	150
—	0.48	.008	60
—	2.8	.010	280
—	1.4	.010	140

[b] Data from Chesselet (*39*). "Salt" values computed by ("dust" × salt/dust).

surface film was inorganic (*35*). Biologically active alkaline earths and alkaline metals are also enriched (*37*), but these effects can be masked in coastal areas by the inclusion of terriginous material in the surface active phase. In different areas, marine biological activity can affect the relative composition of the sea surface and influence the elemental ratios found in a fractionated sea salt derived from that surface. Also, there is a particle size dependent effect in the aerosol (*4*), with the smaller size ranges containing a higher relative proportion of organic material.

The data given by Murozumi *et al.* (*38*) for salt and dust in Arctic and Antarctic snows and ice are of interest. The salt-to-dust ratios in those samples range from 2.5 to 100 and are reasonably within the extremes shown in Table I for atmospheric aerosols. In their samples, the Na/K do not exceed 22 (seawater ratio is 28) and are lowest when the salt-to-dust ratios are low (*39, 40*). The second point is compatible with a salt and dust mixing model (*12*), and the first demonstrates the ratios found in a fractionated marine aerosol (*40*).

Relative Chemical Composition of Marine Aerosol Salt

Since complete analysis of the salt fraction of a marine aerosol is difficult, many investigators have limited their determinations to certain groups of elements such as the halogens or the alkali metals. For these and other reasons, results of such analyses are usually reported as elemental ratios, *e.g.*, I/Cl or Na/K. These ratios may then be compared with the corresponding ratios found for sea water. Meteorological data and air mass histories, which are of primary importance when assessing chemical results, are usually not given, although more recent trends are to include this information (*12, 39–41*).

From inspection of the more recent sea salt aerosol data, it appears that only soluble salt or total aerosols compensated for dust (*39, 40*) were analyzed. These show definite trends in ion enrichments when normalized to Na and compared with sea water ratios. Even with the strict requirements of what the authors consider to be a pure marine sea salt aerosol, the majority of aerosols collected by various workers show elemental ratios different from those of sea water.

Table II shows overall average enrichments (E) selected from the literature for some elements.

$$E = \left[\frac{\text{(element/Na) aerosol}}{\text{(element/Na) sea water}} \right] - 1$$

In the naturally formed sea salt aerosol, Cl, K, Mg, and Ca are almost always enriched with respect to sodium. One reference concerns F, which is depleted in the salt. The reasons for variations in the amount of enrichment are manifold. For instance, although sampling precautions are observed and separations done, it can be seen from the chemical analyses, which correspond to the samples listed in Table I (*39, 40*), that K/Na ratios tend to increase as the dust/salt ratio in the air mass increases. Also, the elements which are enriched in dust are capable of being enriched by the previously mentioned organic sea surface materials—a highly variable component. Although some workers are using F/Cl, F/Na, or Fe/Na ratios in an attempt to monitor terriginous components in the "salt" sample, a routine method to monitor the organic component of the aerosol is also needed. Possibly the method used to determine organic carbon in sea water would suffice.

The common anions (halogens, NO_3^-, SO_4^{2-}, and the organic acids) will or could form volatile compounds in the marine environment. The effects of their chemical activity and physical mobility have held the interest of researchers for many years (*15, 25, 40, 43, 44*). These effects, however, severely limit the usefulness of those elements for chemically defining a marine aerosol or a fractionated sea salt component of that aerosol. In spite of the difficulties and the differences in Table II, Murozumi (*39*) and Chesselet (*40*) are in general agreement as to the overall enrichment of Cl and K in the sea salt component (*45*). This average effect is upheld for K by the previously mentioned data of Murozumi (*38*).

Apparently, enough information about geochemical mass transfer can be gathered to evaluate the elemental contribution of the sea to a marine aerosol. This can be done without actually defining any fractionation processes at the sea surface. However, with the refined sampling techniques and data evaluation considerations required, additional information could be obtained to elucidate the fractionation mechanisms.

Table II. Enrichments of Elements in Sea Salt Aerosols

	Reference			
Element	*40*	*39*	*41*	*42*
F	−0.62	—	—	—
Cl	+0.11	+0.09[a]	—	+0.60[c]
				≥−0.30 ≤0.0[d]
K	—	+0.60[a]	—	+1.0[d]
	+0.40	+0.40[b]	+0.10	—
Mg	—	+0.06[b]	−0.01	+ .17[d]
Ca	—	+0.34[b]	+0.02	+1.7[d]

[a] *In situ* experiments at sea.
[b] Antarctic marine aerosol.
[c] Can include gaseous atmospheric chlorine.
[d] Aerosol with continental component.

Chemical Composition of Marine Aerosol Dust

Our findings for the average elemental concentrations in the same dust samples shown in Table I are presented in Table III as parts per million of dust. For comparison we list the corresponding average elemental concentration for crustal rocks, with which our samples agree. (The details of the table actually show these dusts to be somewhat depleted in Na and K, a feature of sedimentary, not average, crustal rock.) In either case, this is strong evidence for the continental origin of these dusts. Our nondestructive analyses agree to within ±5–30% with other dust analyses reported for the Pacific (*46, 47*) and Atlantic (*22, 38, 48*). Our dust results are compared with two investigations where small aerosol samples were analyzed without salt and dust separation. In both cases, one in the Atlantic (*46*), an area we sampled in a different year, and one in the Indian Ocean (*47*), data are reported as micrograms per cubic meter. If one calculates the ratios element/Fe, for example, one finds values similar to those for our dust, Table IV. Thus, by analyzing for the appropriate transition elements, one can detect the presence of dust in marine aerosols even without separation from the salt.

Relative Chemical Composition of Marine Aerosol Salt vs. *Altitude*

Several workers (*39, 49, 50*) have shown that the atmospheric concentration of salt derived from the ocean surface decreases rapidly to small values at an altitude of about 2 km. Are there also changes in chemical composition in the salt fraction of the marine aerosol with

Table III. Average Amount of Some Elements

Element	*Atlantic*	*Pacific*	*Caribbean*
Al	96,000	—	70,070
Fe	84,000	43,600	44,000
Na	5,400	7,480	7,700
K	—	—	—
Mg	32,000	18,000	16,000
Ti	4,250	15,800	15,200
Mn	717	510	600
F	715	630	544
Cr	—	235	—
V	56	—	—
Ce	86	23	100
Co	43	18	16
Sc	24	15	19
Eu	3	1	1

increasing altitude? We have investigated this for the elements F, Cl, Na, and K. Samples were collected at different altitudes in Hawaii: aboard ship at 7 meters and from a tower on the beach at 14 meters above sea level. Aircraft flights were used to sample at 30 meters on the east coast of the United States, at 300 meters in Puerto Rico, and at 2700 meters at Kodiak, Alaska. The results of the analyses of these samples are given in Figure 2.

Up to an altitude of 300 meters, the relative salt compositions are much like those given in Table II for the ocean-derived salt component. Above 300 meters we see a change in the relative composition to something differing from ocean-derived salt. This might be expected if sea salt is at low concentrations above 2 km. We feel justified in utilizing these data for such a conclusion because even though the samples were

Table IV. Comparison of Elemental Ratios in Aerosols

	Ref. 43	*Ref. 44*	*Dust*
Ce/Fe	1.3×10^{-3}	—	1.02×10^{-3}
Eu/Fe	1.9×10^{-5}	—	3.6×10^{-5}
Sc/Fe	1.8×10^{-4}	—	2.9×10^{-4}
Co/Fe	3.8×10^{-4}	—	5.1×10^{-4}
Mn/Fe	—	4.4×10^{-2}	8.5×10^{-3}
Errors for individual element	$\pm 3\sigma \approx 15\%$	$\pm 3\sigma \approx 15\%$	*See* Table III

collected over different areas, the mechanism of aerosol salt production from the ocean is the same for all areas, and the relative surface sea water composition is almost constant everywhere. Furthermore, weather conditions were quite similar so that we expect comparable total aerosol salt loading in the atmosphere (*49, 50*).

in Over-Ocean "Dust" in Parts Per Million

Greenland Sea	*East Coast, U. S.*	*Crustal Rocks*	*$\pm 1\sigma$ Errors, %*
—	—	81,300	5–20
78,000	50,000	50,000	1–5
17,500	—	28,300	10–50
9,250	—	25,900	15
—	—	20,900	5–25
—	—	4,400	20–75
970	—	1,000	10–99
875	—	700	2–5
730	100	200	4–15
—	—	110	15–20
60	—	46	2–60
46	50	23	2–25
27	13	5	1–10
1	—	1	10–30

The figure also includes land breeze data from two different altitudes, 90 and 1800 meters, which were obtained within 2 hours of each other from a circling aircraft. Again we see deviations from sea water ratios with increasing altitude for Cl/Na. The absolute values of the ratios containing F also show this change, but are different from marine aerosol ratios. In this case, however, the wind was from the land and the samples were collected about 150 miles from the U. S. east coast. The influence of a continental aerosol component did appear in the fluoride concentration (*12*). Similarly, changes in the relative chemical composition of marine aerosols with altitude have been reported by others for SO_4/Cl (*2, 51*) and Cl, Br, and I (*44*).

Relative Chemical Composition of Oceanic Rain and Snow

In many instances the chemical composition of oceanic rain has been used to draw conclusions concerning the chemical composition of marine aerosols (*3, 52, 53, 54*). The first two entries in Table V emphasize the trends with respect to altitude for our data on rain samples collected in Hawaii. The 100-meter sample was collected near the coast, and the 1000-meter sample was obtained several miles inland in the Hawaiian mountains. The results for F/Cl, Cl/Na, and Na/F in the 100-meter sample agree with the results found for marine aerosol salt fractions given in Table II, and differ only slightly from those of sea water.

In the 1000-meter sample, the total salt concentration in the rain decreased by a factor of almost 20. In this sample, the elemental ratios are quite different from sea water ratios. Except for Cl/Na, they resemble the aerosol salt ratios shown in Figure 2 for altitudes above 2 km and for samples off the east coast in which continental influences were present. Further interpretation of such data is complex as a thorough investigation of rain phenomena in Hawaii has shown (*55*). Also included in the table are data for fog collected in the Arctic Ocean region. In these condensates, chloride is almost always enriched relative to sodium when compared to sea water. Chloride is more enriched where the total sodium content is less and with higher altitudes.

Only tentative conclusions may be drawn from rain data concerning aerosols. The interactive processes between rain and aerosols such as wash out, rain out, and precipitation scavenging are not well understood. It has been pointed out by Wilkniss and Bressan (*40*), for example, that elemental ratios in low Na content rain may be the result of the rain-forming process rather than being representative of the aerosols present in the air parcel. Indeed, better aerosol data will assist the description of the rain-forming processes rather than vice-versa.

Of great interest and value have been analyses of snow-formed ice in the polar regions. In these ice sheets, we find preserved records of aerosols from the marine atmosphere. Recent work on lead in Greenland and Antarctic snow strata (*38*) and on sulfate and mercury also in Greenland ice and snow (*56–58*), clearly show increases in these constituents in recent decades which might be attributable to man-made pollution.

Table V. Rain Collected in the Marine Environment

Location	*Altitude, meters*	*F/Cl* ($\times 10^5$)	*Na/K*	*Cl/Na*	*Na/F* ($\times 10^{-3}$)	*Total Na in Rain, μgram/ml*
Hawaii	100	9	—	1.8	6.2	12.2
	1000	164	—	3.4	0.2	0.7
	3	40	—	2.2	1.1	21
	100	44	—	1.8	1.2	3.9
	1000	108	—	2.8	.3	1.0
	1000	62	—	3.8	.4	0.8
Arctic Fog 1	6	8.5	12.6	2.2	5.3	75
Arctic Fog 2	6	178	16.0	1.7	0.3	2.3
Arctic Fog 3	6	36	14.7	2.4	1.1	9.5

Discussion

Inorganic aerosols near the ocean–atmosphere interface are composed of both salt and dust. The source of the dust is the continents. This is indicated by analyses that show the elements present in proportions close to those found for average crustal rocks. Further evidence is obtained from mineralogical investigation. Dust over the equatorial Atlantic has been found to originate in Africa, and dust in the South Pacific was traced to Australia. Little is known about the chemistry of aerosol dusts over the oceans. Significant changes in the dust component may occur

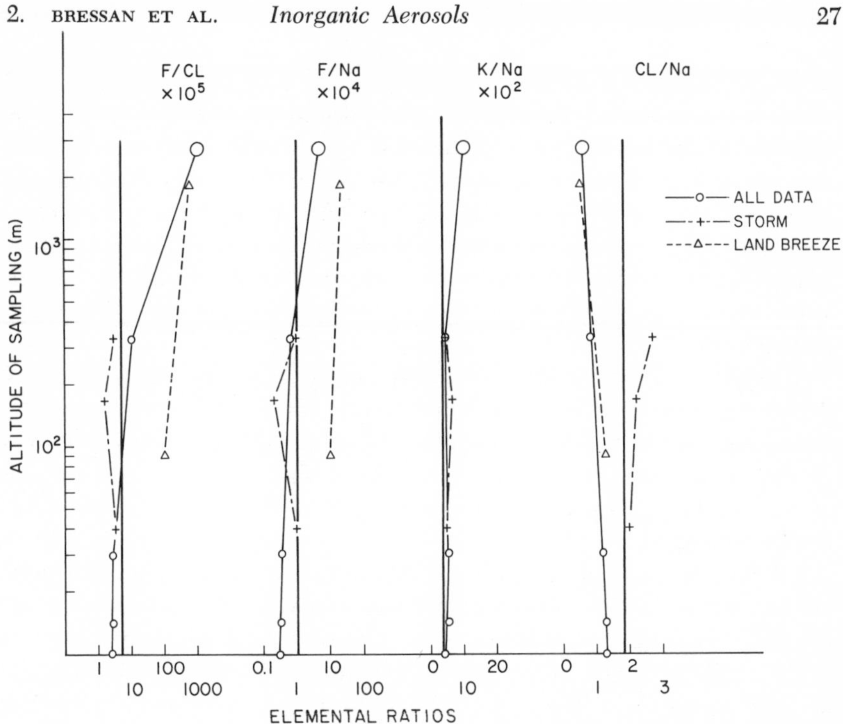

Figure 2. Elemental ratios in marine aerosols vs. *altitude. Width of symbols indicates* $\pm 2\sigma$ *error limits.*

when volcanic eruptions add ash and other emanations such as sulfuric acid (*59, 60*). Usually these heavy volcanic contributions are of a short-lived nature near the ocean–atmosphere interface. If the eruption cloud penetrates into the atmosphere, tropospheric aerosols may be altered at distant times and places (*61*). Leaching of dusts may occur if the dust particles serve as condensation surfaces for atmospheric moisture. However, subtle changes in the trace element chemistry of the original dust cannot be detected at the ocean–atmosphere interface since the dust is in contact with salt during sampling and then washed with water prior to analysis.

The source of the aerosol salt is the ocean. The elemental constituents occur in ratios which differ somewhat from those found in sea water. The chemical composition of the aerosol salt is influenced by several factors. When the salt aerosol is created at the sea surface by whitecaps, we find ion fractionation. That is, the sea water droplets injected into the atmosphere contain the elements in proportions different from those in sea water. Examples include different ratios for the halogens, alkali metals, sulfate, phosphate, and nitrogen (*12, 44, 51, 62, 63*). The causes for fractionation, both as physical and organic chemical processes are under study (*4, 52*). The aerosol salt composition may be further modified when the droplets are carried to higher altitudes where they lose water by evaporation, crystallize, and then lose other constituents such as chlorine gas (*12, 26, 64*). Another possibility of change in the chemical composition is the reaction of the aerosol salt with atmospheric gases such as elemental iodine, which originates at the ocean surface (*65*). In addition, soluble particles are produced from gaseous compounds in the marine atmosphere, such as the production of $(NH_4)_2SO_4$ *via* the atmospheric oxidation of SO_2 (*9*).

Salt-to-dust ratios may vary considerably over the different oceans, depending on source areas and wind transport. The salt component predominates within the first kilometer above the ocean surface, then decreases rapidly to very small values at 2 km. Above 2 km, sea salt aerosols appear to be insignificant when compared with the continental dust component (*2, 3, 32, 66*).

Finally man-made aerosols contribute to natural marine aerosols in coastal areas. In the marine atmosphere, however, this pollution is still negligible when compared with natural aerosol production (*67, 68*).

A working symposium on sea–air chemistry (WORKSSAC) was held at Fort Lauderdale, Fla., and papers presented there were published (*69*). These include the most recent work in the field. Recommendations were also made at the symposium for areas of research and conventions for data presentation. Because this manuscript was compiled before the new conventions were adopted, they have not been observed here. However, the formula for enrichment (E) is in accordance with those conventions.

Acknowledgment

We are indebted to T. B. Warner of the Naval Research Laboratory who performed the fluoride analyses of sea water, sea spray, and rainwater and collected rain samples in Hawaii. Technical assistance during various analyses was rendered by Elwood Russ, who, along with Gene Bugg, helped collect kite flight and aircraft samples. Robert Beckett of the Marine Biology Branch, Naval Research Laboratory, took the electron micrographs.

Literature Cited

1. Bruyevich, W. V., Kulik, Ye. Z., *Dokl. Akad. Nauk. SSSR* (1967) **175,** 697.
2. Junge, C. E., Robinson, E., Ludwig, F. L., *J. Appl. Meteorol.* (1969) **8,** 340.
3. Junge, C. E., *Geochim. Cosmochim. Acta* (1968) **32,** 1219.
4. Barker, D. R., Zeitlin, H., *J. Geophys. Res.* (1972) **77,** 5076.
5. Twomey, S., Wojciechowski, T. A., *J. Atmos. Sci.* (1969) **26,** 648.

6. Windom, H. H., *Geochim. Cosmochim. Acta* (1970) **34,** 509.
7. Ninkovich, D., Heezen, B. C., *Nature (London)* (1967) **213,** 582.
8. Horn, D. R., Delach, M. N., Horn, B. M., *Geol. Soc. Amer. Bull.* (1969) **80,** 1715.
9. Dinger, J. E., Howell, H. B., Wojciechowski, T. A., *J. Atmos. Sci.* (1970) **27,** 791.
10. Parkin, D. W., Tilles, D., *Science* (1968) **159,** 936.
11. Chow, T. J., Earl, J. L., Bennett, C. F., *Environ. Sci. Technol.* (1969) **3,** 737.
12. Wilkniss, P. E., Bressan, D. J., *J. Geophys. Res.* (1971) **76,** 736.
13. Prospero, J. M., Bonatti, E., *J. Geophys. Res.* (1969) **74,** 3362.
14. Parkin, D. W., Phillips, D. R., Sullivan, R. A. L., *J. Geophys. Res.* (1970) **75,** 1782.
15. Duce, R. A., Winchester, J. W., Van Nahl, T. W., *J. Geophys. Res.* (1965) **70,** 1775.
16. Jacobs, W. C., *Month. Weath. Rev.* (1937) **65,** 147.
17. Mitchell, R. I., Pilcher, J. M., *Ind. Eng. Chem.* (1959) **59,** 1039.
18. Nan-Hai Hu, J., *Environ. Sci. Technol.* (1971) **5,** 251.
19. Barger, W. E., Garrett, W. D., *J. Geophys. Res.* (1970) **75,** 4561.
20. Shedlovsky, J. P., Paisley, S., *Tellus* (1966) **18,** 2, 499.
21. Dixon, B. E., Metson, P., *Analyst (London)* (1960) **85,** 123.
22. Squires, P., Twomey, S., *J. Atmos. Sci.* (1966) **23,** 401.
23. Twomey, S., *J. Atmos. Sci.* (1971) **28,** 377.
24. Gatz, D. F., Dingle, A. N., *Tellus* (1971) **23,** 14.
25. Eriksson, E., *Tellus* (1960) **12,** 63.
26. Duce, R. A., *J. Geophys. Res.* (1969) **74,** 4599.
27. Junge, C. E., "Air Chemistry and Radioactivity," 1963, Academic Press, New York.
28. Wilkniss, P. E., Larson, R. E., "Use of Activation Analysis to Determine the Chemical Composition and Origin of Particles Collected in the Marine Atmosphere," Proceedings of the International Atomic Energy Agency Symposium on Uses of Nuclear Techniques in the Measurement and Control of Environmental Pollution, IAEA/SM 142/6, Salzburg, Austria, 1970.
29. Warner, T. B., *Science* (1969) **165,** 178.
30. Perkin-Elmer Corp., "Analytical Methods for Atomic Absorption Spectrophotometry," 1971, Norwalk, Conn.
31. Roll, U. H., "Physics of the Marine Atmosphere," 1965, Academic Press, New York.
32. Chester, R., Elderfield, H., Griffin, J. J., *Nature (London)* (1971) **233,** 474.
33. Woodcock, A. H., *J. Meteorol.* (1953) **10,** 362.
34. Blanchard, D. C., Syzdek, L. D., *J. Geophys. Res.* (1972) **77,** 5087.
35. Piotrowicz, S. R., Ray, B. J., Hoffman, G. L., Duce, R. A., *J. Geophys. Res.* (1972) **77,** 5243.
36. Wallace, G. T. Jr., Loeb, G. I., Wilson, D. F., *J. Geophys. Res.* (1972) **77,** 5293.
37. Szekielda, K. H., Kupferman, S. L., Klemas, V., Polis, D. F., *J. Geophys. Res.* (1972) **77,** 5278.
38. Murozumi, M., Chow, T. J., Patterson, C., *Geochim. Cosmochim. Acta* (1969) **33,** 1247.
39. Chesselet, R., Morelli, J., Buat-Menard, P., *J. Geophys. Res.* (1972) **77,** 5116.
40. Wilkniss, P. E., Bressan, D. J., *J. Geophys. Res.* (1972) **77,** 5307.
41. Hoffman, G. L., *J. Geophys. Res.* (1972) **77,** 5161.
42. Tsunogai, S., Saito, O., Yamada, K., Nakaya, S., *J. Geophys. Res.* (1972) **77,** 5283.
43. Green, W. D., *J. Geophys. Res.* (1972) **77,** 5152.

44. Winchester, J. W., Duce, R. A., *Naturwissenshaften* (1967) **54,** 110.
45. Chesselet, R., personal communication, 1972.
46. Dudley, N. D., Ross, L. E., Noshkin, V. E., "Application of Activation Analysis and Ge(Li) Detection Techniques for the Determination of Stable Elements in Marine Aerosols," 1968, p. 55, Proceedings of the International Conference on Modern Trends in Actication Analysis, National Bureau of Standards, Gaithersburg, Md.
47. Egorov, V. V., Zhigalovskaya, T. N., Malakhov, S. G., *J. Geophys. Res.* (1970) **75,** 3650.
48. Chester, R., Johnson, L. R., *Nature (London)* (1971) **231,** 176.
49. Eriksson, E., *Tellus* (1959) **11,** 375.
50. Hobbs, P. V., *Quart. J. Roy. Meteorol. Soc.* (1971) **97,** 263.
51. Abel, N., Jaenicke, R., Junge, C., Kanter, H., Rodriques, G., Pietro, P., Seiler, W., *Meteorol. RdSch.* (1969) **22,** 158.
52. Horne, R. A., "Marine Chemistry," 1969, Wiley-Intersciene, New York.
53. Lazrus, A. L., Baynton, H. W., Lodge, J. R., *Tellus* (1970) **22,** 106.
54. Seto, Y., Duce, R. A., Woodcock, A. H., *J. Geophys. Res.* (1969) **74,** 1101.
55. Kruger, P., Miller, A., *J. Geophys. Res.* (1966) **71,** 4243.
56. Federer, B., *Pure Appl. Geophys.* (1970) **79,** 120.
57. Koide, M., Goldberg, E. D., *J. Geophys. Res.* (1971) **76,** 6589.
58. Weiss, H. V., Koide, M., Goldberg, E. D., *Science* (1971) **174,** 692.
59. Cadle, R. D., Wartburg, A. F., Grahek, F. E., *Geochim. Cosmochim. Acta* (1971) **35,** 503.
60. Cadle, R. D., Blifford, I. H., *Nature (London)* (1971) **230,** 573.
61. Lamb, H. H., *Phil. Trans. Roy. Soc.* (1970) **266,** 425.
62. Bloch, M. R., Luecke, W., *Israel J. Earth Sci.* (1970) **19,** 41.
63. Bloch, M. R., Luecke, W., *J. Geophys. Res.* (1972) **77,** 5100.
64. Sugawara, K., *Oceanogr. Mar. Biol. Annu. Rev.* (1965) **3,** 59.
65. Martens, C. S., Harris, R. C., "Precipitation Scavenging," *USAEC Symp.* (1970) **22,** CONT-700601.
66. Junge, C. E., *J. Geophys. Res.* (1972) **77,** 5183.
67. Landsburg, H. E., *Science* (1970) **170,** 1265.
68. Ellis, H. T., Pueschel, R. F., *Science* (1971) **172,** 845.
69. Duce, R. A., Stumm, W., Prospero, J. M., *J. Geophys. Res.* (1972) **77,** 5059.

RECEIVED January 7, 1972.

3

The Sources and Distribution of Vanadium in the Atmosphere

W. H. ZOLLER, G. E. GORDON, E. S. GLADNEY, and A. G. JONES[1]

Department of Chemistry, University of Maryland, College Park, Md. 20742

With modern methods of analysis of atmospheric particulate matter, especially neutron activation, one can determine atmospheric concentrations of vanadium in remote locations. Comparisons of vanadium concentrations with those of other elements in remote areas suggest that much of the observed vanadium orignates from man's activities. Atmospheric vanadium concentrations in most United States cities are less than 20 ngrams/meter^3, but many cities in the northeastern United States have up to several μgrams/meter^3. A study of Boston shows that residual fuel combustion is the only source of vanadium of sufficient magnitude to produce the concentrations observed. Because of the high sensitivity for its analysis, vanadium can serve as an indicator of wide-scale movement of particulates from this identified anthropogenic source.

Accurate measurements of the concentrations of trace elements on atmospheric particulates are difficult enough to make in polluted urban atmospheres but even more so in "clean" marine or polar atmospheres because of the minute quantities of material that can be collected in a reasonable time. For these measurements, one needs a sensitive analytical technique that is free from interference by other elements present. Recently, the use of lithium-drifted germanium [Ge(Li)] γ-ray detectors in neutron activation analysis has greatly improved analytical sensitivities and accuracies for such studies (*1*).

One element that can be measured by this method is vanadium (V). In most atmospheric particulate samples (as well as important pollution

[1] Present address: Shields Warren Radiation Laboratory, Harvard Medical School, Boston, Mass.

source materials such as oil and coal), vanadium can be measured simply by irradiating the samples with neutrons and counting the γ-rays emitted by the neutron-capture product, 3.8-min ^{52}V, with Ge(Li) detectors without any prior chemical separations. In some regions, particularly marine environments, γ-rays from sodium (Na) and chlorine (Cl) activities may obscure those of ^{52}V, making it necessary to perform a simple chemical step before counting (*2, 3*).

Vanadium is not known to be a serious hazard to human or other biological life (although it may have an undesirable catalytic effect on oxidation of SO_2). Nevertheless, as discussed below, vanadium may be useful as an indicator of the global movement of particulates from man's activities.

Atmospheric Concentrations of Vanadium in Remote Areas

Consider the atmospheric concentrations of vanadium in areas as remote as possible from man's activities. Few data of this sort exist, but they are sufficient to suggest trends that deserve further investigation. In Table I, we summarize the available data on vanadium from remote areas of the Pacific Ocean and Canada. The small concentrations observed, down to 0.1 ngram/meter^3, illustrate the difficulty of the measurements noted above. By contrast, in some cities, vanadium concentrations are sometimes a few μgrams/meter^3 (*see* Table V), ten thousand times greater than in the Eastern Pacific.

Consider whether the natural origins of vanadium account for the amounts observed. The major natural sources are marine aerosols produced by the bursting of bubbles at the sea surface and continental dust from wind erosion of rocks and soil. Some vanadium would also be injected into the atmosphere by volcanoes, but in the treatment below this is a minor perturbation on the dust contribution.

One could perhaps predict the amounts of airborne vanadium from these sources from estimates of the total amount of particles injected by the various natural processes multiplied by the typical vanadium concentration in the materials injected. However, present knowledge of the former is highly uncertain, and one would need considerable additional information on global transport rates and the residence times of V-bearing particles to estimate atmospheric concentrations at the given locations.

There is, fortunately, an alternative means of prediction that yields rather definitive results: one can predict V concentrations relative to those of observed elements known to arise from those sources. For example, in a marine atmosphere, most of the airborne Na and Cl is present on sea-salt particles. Thus, to estimate the upper limit of the marine contribution of V in a given location, one can multiply the observed Na and Cl concentrations by the V/Na and V/Cl ratios of sea

Table I. Atmospheric Vanadium Concentrations of Remote Nonurban Regions

Area	*No. of Samples*	*V Concn, (ng/meter³)* *Range*	*Av Concn*	*Ref.*
Eastern Pacific	9 1-day samples	<0.02–0.8	0.1	*3*
Windward Hawaii	3 1-day samples	0.1–0.3	0.2	*3*
Windward Hawaii	9 1-week samples	0.14–1.0	0.34	*2*
Canada	24 2-week samples at 5 sites	0.21–1.9	0.72	*4*

water, respectively, assuming that there is no mechanism for concentrating V more strongly than those elements on marine aerosols. Likewise, we can obtain an upper limit of the V contribution from continental dust and volcanoes by assuming that all of the observed iron (Fe) arises from those sources and multiplying the Fe concentration by the V/Fe ratio for those source materials. One need not restrict the predictions to those elements, but can double check them by using ratios with additional elements such as Mg in sea water and Al, Sc, La, etc., in continental dust and volcanic debris.

Table II lists the concentration ratios of V to several other elements for the natural sources considered. For continental dust and volcanic material, there are several possible choices of V ratios: crustal abundances, sediments, average soils, or, in the case of volcanic material, basic rocks. Fortunately, however, the ratios of V concentrations to those of the other elements considered vary little among these choices.

In Table III we show the concentrations of vanadium predicted for windward Hawaii relative to Na and Mg for the marine aerosol component and to Fe, Mn, and Al for continental dust based on the excellent data of Hoffman (*2*). We have chosen this particular set of Pacific area data for analysis because it includes careful analyses for the several elements in addition to vanadium needed for the predictions. In Table III we see that the marine V concentrations based on Na and Mg agree, showing that the Mg/Na ratio on aerosols collected there is the same as the sea water value. The V contribution from marine aerosols is quite small compared with both the observed concentration and the amount predicted from continental dust. The latter value is based on the elemental ratios for soils in Table II. Predictions based on Fe, Al, and Mn are in reasonable agreement with each other, a result of the fact that the ratios Fe/Al/Mn for windward Hawaii aerosols are about the same as for the soil component of Table II.

Finally, the most important result in Table III is that the predicted V concentrations from natural sources account for only 31% of the observed amount. Although it may be a coincidence, it is interesting that the predicted V concentration from continental dust, which would

Table II. Concentrations of Vanadium Relative to Those of Other Elements in Natural Sources

A. Continental Dust and Volcanic Debris

Element	Concentration Ratio			
	Diabase [a]	*Crustal Av* [a]	*Soils* [b]	*Sedimentary Rocks* [b]
V/Fe	0.0031	0.0027	0.0026	0.0039
V/Mn	0.18	0.14	0.12	0.19
V/Al	0.0031	0.0017	0.0014	0.0018
V/Zn	2.9	1.9	2.0	1.6
V/Sc	7.1	6.1	14.3	13
V/Co	4.8	5.4	12.5	5.6
V/Sb	220	676	100	100
V/La	8	4.5	2.5	3.2

B. Marine Aerosols

Element	*Sea Water* [a]
V/Na	2×10^{-7}
V/Mg	1.5×10^{-6}
V/Cl	10^{-7}

[a] Mason (*5*).
[b] Vinogradov (*6*).

be in the form of mineral particles, agrees exactly with the observed amount of acid-insoluble V as observed by Hoffman (*2*). It is tempting to suggest that the remaining vanadium, which apparently arises from man's activities, is present in an acid-soluble form, *e.g.*, an oxide or sulfate.

One should probably use elemental ratios for types of rocks more basic than soil (*e.g.*, diabase) to account for the small expected volcanic component. Although we have not used them in the calculations, we see in Table II that for many of the elements listed, especially Fe and Mn, the relative V concentrations do not vary greatly from one rock or soil type to another; thus, the predictions are not very sensitive to one's choice of a basis for the predictions. In particular, any small volcanic component is included implicitly in calculations based on Fe and Mn and, to a lesser extent, Al. Thus, we must conclude that about two-thirds of the vanadium observed at the windward Hawaii site must arise from man's activities, unless there is some mechanism by which vanadium is enriched relative to the other elements when natural aerosols are formed.

A similar calculation may be done for Rahn's data (*4*) for rural Canada as shown in Table IV. As would be expected for a midcontinental area, the marine component of vanadium is quite small. The variations in predicted V from continental dust based on the elements Fe, Mn, and

Al are greater than for the windward Hawaii data. The variations continue to be large when the analysis is extended to Sc and La. These variations simply reflect the fact that the relative concentrations of these five elements on Canadian aerosols are not the same as those of the average soils as computed by Vinogradov (*6*). There are many possible reasons for these variations, *e.g.*, the rocks and soil in the source area may have compositions different from the world-average soil, and there may be some man-made aerosols even in this remote area. We see the latter effect particularly for the vanadium concentrations based on Zn and Sb, both of which have relative concentrations about 30-fold greater than the soil average. It seems likely that the large enrichments for these elements result from man's activities. Rahn (*4*) has discussed these observations in more detail.

Because of the variability of the predictions of the V contribution from continental dust, we are unable to determine whether or not the observed V is fully accounted for by natural sources. The prediction in Table IV agrees perfectly with the observed concentration, but the uncertainty of the prediction (at least ± 0.4 ngram/meter^3) and the variation of the observed V concentration from one sampling site to another in Canada are easily large enough to conceal an unexplained V component of 0.24 ngram/meter^3 found in our analysis of the windward Hawaii data. Thus, although it seems clear from the Hawaii data that there is a widespread, perhaps global, contribution of vanadium from man's activities, we need additional data from remote areas where natural V concentrations are small to find additional evidence for atmospheric

Table III. Predicted Vanadium Concentrations for Windward Hawaii from Natural Sources

Source	*Element used as Basis*	*Concn of Element,[a] ng/meter^3*	*Predicted V Concn, ng/meter^3*	
Marine aerosols	Na	5300	0.0010	0.0010
	Mg	670	0.0010	
Continental dust	Fe	38	0.099	0.114
	Mn	0.86	0.103	
	Al	101	0.141	
			Total	0.11

Observed[a]	Acid soluble	0.24 ± 0.10
	Insoluble	0.11 ± 0.08
	Total	0.35 ± 0.13
Predicted natural/observed		31%

[a] Hoffman (*2*).

Table IV. Predicted Vanadium Concentrations for Rural Canada from Natural Sources

Source	*Element used as Basis*	*Concn of Element,[a] ng/meter³*	*Predicted V Concn, ng/meter³*	
Marine aerosols	Na	44	0.0000085	0.000005
	Cl	13	0.0000019	
Continental dust	Fe	210	0.55	0.73
	Mn	6.7	0.80	
	Al	186	0.26	
	Sc	0.11	1.57	
	La	0.20	0.50	
	Zn	15.4	31	Anomolous
	Sb	0.25	25	
			Total	≈0.73

Observed[a] 0.72 ± 0.50
Predicted natural/observed ≈100%

[a] Rahn (*4*) Average of five winter stations.

vanadium from pollution sources. Gladney *et al.* (7) recently initiated collection and analyses of particulate samples from Antarctica, which probably has the cleanest atmosphere that can be found at ground level on earth. Indeed, vanadium concentrations down to 0.0007 ngrams/meter³ were reported from preliminary analyses of their samples, more than 100-fold smaller than the values found in the relatively "clean" atmosphere of Hawaii and the northern Pacific (Table I). As yet the samples have not been analyzed for enough other elements to permit the kind of analysis shown in Tables III and IV for other areas.

Urban Atmospheric Vanadium Concentrations

The windward Hawaii data suggest that even in an area remote from major human activities, about two-thirds of the atmospheric vanadium originates from anthropogenic sources. In an attempt to identify the sources, let us consider the vanadium concentrations in the atmospheres of a number of U. S. cities, as listed in Table V. As shown in the table, U. S. cities can clearly be separated into two distinct groups: group A, characterized by V concentrations of an average of about 11 ngrams/meter³ (approximately 20 times that of remote areas), and group B, in which overall V concentrations range between 150 and 1400 ngrams/meter³. Although there is considerable disagreement among some of the different concentration values reported for particular cities (arising from different locations and conditions during sampling and possible analytical errors), the V concentrations are clearly one or two orders of magnitude greater than those of the group A cities. What causes the big difference in vanadium concentration among the two groups of cities? All of the group B cities are in the northeastern United States. Although the number of cities examined here is too small to obtain valid statistics, this conclusion is supported by the values reported for many more cities in the NASN compilation (*14*). Of the 57 urban areas for which vanadium concentrations are listed in the compilation, all but three of the 14 stations reporting average V concentrations above 30 ngrams/meter^3 are east of Ohio and north of Washington, D. C. The most important difference in air pollution sources in the two groups of cities is the fact that in the group B cities of the Northeast, large amounts of residual oil (also called No. 6 oil or "Bunker-C") are burned for heat and electric power generation. The major source of this residual oil is Venezuelan crude oil which yields residuals typically having V concentrations in the range of 200–1000 ppm.

Table VI shows the results of predictions similar to those of Tables III and IV for several of the cities for which V concentrations are given in Table V. We have not shown the marine aerosol components since their V contributions are negligible compared with the continental dust components.

One interesting observation from Table VI is the fact that for the group A areas, we account for 40–100% of the observed atmospheric vanadium with the use of the continental dust component alone, despite the fact that there are many other local sources of particulates in each area. This result is in part fortuitous, because the V concentrations relative to elements such as Fe, Mn, and, to a lesser extent, Al are about the same in some pollution sources (*e.g.*, coal fly ash) as in the soil average used for the continental dust component (see Table VII). It is particularly interesting that the vanadium concentrations in the northwest Indiana and Chicago area are comparable with those of other group A urban areas although the former is one of the world's greatest steel-making areas, and vanadium is associated with other ferrous metal ores as well as being an alloying metal in some kinds of steels. It would thus appear that iron-ore processing and steel making are not important contributors to the worldwide atmospheric vanadium concentrations.

The differences in observed V concentrations between the latter two sets of data for New York City in Table V probably arise in part from differences in sampling locations: both were area-wide surveys, but that by Kneip *et al.* (*16*) was confined to Manhattan Island, where the air pollutants are probably most heavily concentrated whereas that by Morrow and Brief (*15*) covered a much greater area around the city. Vanadium concentrations obtained by the NASN (*14*) by emission spectrometry are frequently lower by a factor of 2 or 3 than those ob-

Table V. Atmospheric Vanadium Concentrations of Urban Regions

Area	No. of Samples	V Concn, ng/meter³ Range	Av Concn	Ref.
	Group A			
Honolulu, Hawaii	12	0.68–2.4	3.4	2
Los Angeles (UCLA)	18	2.2–28	12	2
San Francisco	9	1.5–11	6.2	8
Niles, Michigan	25	1.2–16	4.5	9
Northwest Indiana	25	5–18	7.4	10
Chicago	22	2–120	22	11
Urban England	—	6–60	21	12
	Group B			
Buffalo, N. Y.	6	500–2000	800	13
New York City	25	74–2000	458	14
New York City	270	≈90–320	170	15
New York City	52	310–2790	1,320	16
Boston	—	70–230	160	14
Boston	10	400–2000	600	1
Boston	12	90–1320	494	17
Boston	9	89–2410	940	18

tained by more modern techniques (*see,* for example, the data for Boston and New York in Table V). We do not know whether this discrepancy exists because emission spectroscopy is less accurate and sensitive than techniques such as activation analysis and atomic absorption used in the more recent studies, or if the vanadium concentrations have increased considerably between the early 1960's when most of the NASN results were obtained and the late 1960's and early 1970's when the other data were taken.

Regardless of uncertainties in the values of the V concentrations, it is clear that in group B cities such as New York, we are able to account for only a few percent of the observed vanadium by the method that was rather successful in predicting concentrations of the group A cities. To fit the concentrations observed in the group B cities, one must include a component in which the ratios of V to Fe, Al, Mn, and other elements are much greater than those of soils, crustal rocks, fly ash from coal burning, etc. As suggested, we feel that the only reasonable V source of the magnitude required is the combustion of residual oils in the group B cities.

Now let us more carefully examine this hypothesis for the origin of vanadium in the group B cities using Boston as an example. Table VII shows vanadium concentrations relative to those of several other elements in several source materials. We have included data for crude oils from several areas to illustrate the enormous variability of elemental

Table VI. Predicted Vanadium Concentrations for Several U. S. Urban Areas, Continental Dust Component Only

Element used as Basis	*Concn of Element, μg/meter³*	*Predicted V Concn, ng/meter³*	
Honolulu, University of Hawaii Site (12 Samples)[a]			
Fe	0.52	1.35	1.64
Mn	0.014	1.65	
Al	0.73	1.92	
Observed	Acid-soluble	2.6 ± 2.2	
	Insoluble	0.83 ± 0.36	
	Total	3.4 ± 2.3	
	Predicted/observed	47%	
Los Angeles, UCLA Site (18 Samples)[a]			
Fe	1.5	3.9	5.2
Mn	0.0053	6.3	
Al	3.8	5.3	
Observed	Acid-soluble	9.7 ± 7.6	
	Insoluble	2.8 ± 2.5	
	Total	12.5 ± 8	
	Predicted/observed	42%	
San Francisco, Area-Wide Study (9 Samples)[b]			
Fe	1.92	5.0	2.9
Mn	0.019	2.2	
Al	0.99	1.4	
	Observed	6.2 ± 2.1	
	Predicted/observed	47%	
Northwest Indiana, Area-Wide Survey (25 Samples)[c]			
Fe	3.9	10.1	9.4
Mn	0.13	15.3	
Al	1.95	2.7	
	Observed	8.2 ± 2.8	
	Predicted/observed	114%	
New York City, Area-Wide Survey (≈270 Samples)[d]			
Fe	2.98	7.4	3.9
Al	2.04	2.8	
Si	4.84	1.5	
	Observed	170	
	Predicted/observed	2%	
New York City, Area-Wide Survey (≈150 Samples)[e]			
Mn	0.071	8.4	20
Cr	0.063	31	
	Observed	1190	
	Predicted/observed	2%	

[a] Hoffman (*2*).
[b] John *et al.* (*8*).
[c] Harrison *et al.* (*10*).
[d] Morrow and Brief (*15*).
[e] Kneip *et al.* (*16*).

ratios in the various crudes. Data in the last column of Table VII were obtained from preliminary analyses of several samples of the residual oils burned in Boston. Since most of the residual oil consumed in the Northeast is obtained from Venezuelan crudes, one would expect the ratios for the former to be more similar to those of the latter than to those of other crudes. However, we see that V is enriched relative to all of the other elements when residual oil is obtained from the crudes. Vanadium is largely present as an organometallic porphyrin compound and is not removed with other metals in the mineral impurities of the crudes. Furthermore, the vanadium compounds have such low volatilities that they remain in the residuum during distillation, becoming more concentrated than in the original crude. The Venezuelan crude studied by Filby and Shah (*21*) contained 112 ppm V, whereas the average V concentration of Boston area residuals is 870 ppm (*22*).

In Table VIII we predict the vanadium concentrations in the Boston area from various sources. In the top part of the table we calculate upper limits of the vanadium contributions from continental dust and coal by in each case assuming that the entire observed concentrations of the elements used as bases result from the indicated source. We see that neither of these sources yields more than a few percent of the observed average V concentration, 600 ngram/meter^3. In the lower part of the table, we predict the V contributions from residual oil relative to several elements. Since the analyses of Boston-area fuels are preliminary, we have also used ratios for Venezuelan crudes for the predictions. We cannot use Fe, Mn, and Sc for these predictions, because the ratios of V to those elements in oil are so great that one obtains vast predicted V concentrations (up to 1700 μgram/meter^3) if one assumes that those elements result entirely from that source. This result simply indicates that little of the atmospheric concentration of those elements is produced by oil combustion. The predictions based on Co, Zn, Sb, and Se show that one can easily obtain the observed amount of atmospheric vanadium from residual oil if only a very small fraction of any one of these elements results from oil combustion. The great magnitude of vanadium released by residual oil combustion is further indicated by the data of Table IX which summarizes the uses of coal and residual oil in the Boston area. We see that there is more than 300 times as much vanadium contained in the residual oil than in the coal burned in Boston. Furthermore, most of the vanadium in oil is released to the atmosphere, whereas much of that in coal is contained in large silicate particles that never leave the stack.

Residual oil combustion can easily account for the observed vanadium concentration in the Boston atmosphere. We have not considered some other sources of particulates in the Boston area, notably refuse

Table VII. Relative Vanadium Concentrations in Soils and Several Pollution Sources

			Relative Concentration				
Elements	*Soils*[a]	*Coal (W. Va)*[b]	*Crude Oils*				*Residual Oil*[e] *(Boston Supply)*
			Calif.[c]	*Libya*[c]	*Wyo.*[c]	*Venezuela*[d]	
V/Fe	0.0026	0.0025	1.3	5.1	52	3.0	1240
V/Mn	0.12	0.78	81	16.4	330	530	8700
V/Al	0.0014	0.0018	—	—	—	—	—
V/Zn	2.0	0.67	9.4	1.3	44	43	1450
V/Sb	100	—	1780	480	4200	406	1.74×10^4
V/Sc	14.3	—	7.9×10^4	6.1×10^4	3.4×10^5	2.5×10^4	8.7×10^5
V/Se	10,000	—	106	85	930	304	2900
V/Co	12.5	3.6	83	268	3300	565	2900

[a] Vinogradov (*6*).
[b] Abernethy *et al.* (*19*).
[c] Filby *et al.* (*20*).
[d] Filby and Shah (*21*).
[e] Gladney and Baumgartner (*22*).

disposal by open burning or municipal incineration and manufacturing. The amount of particulate materials released by industrial operations in the Boston area is so small that they could not possibly be a major vanadium source (*23*). Furthermore, it is difficult to conceive of any sizeable amounts of material in urban trash that would have V/Fe ratios appreciably greater than those observed in the steel-making regions of Chicago and northwest Indiana.

Additional evidence for the origin of vanadium in the Boston atmosphere is shown by the size distributions of vanadium-bearing particles as observed in samples collected with cascade impactors (*18*). In Figure 1 we plot the weight of vanadium per cubic meter of air *vs.* the stage designation of a six-stage cascade impactor. Stage A at the right collects the largest particles and Stage F the smallest. Data are shown for three sites, the first two on the M.I.T. campus in Cambridge and the third near circumferential highway, Route 128, in Wellesley. We see that the vanadium is preferentially associated with the smallest particles collected, as would be expected for a high-temperature process such as oil combustion followed by condensation of the vaporized material. The higher concentrations for the two M.I.T. sites show that the vanadium sources are most likely located in town. We do not have an explanation for the secondary peak consistently observed at intermediate stages of the impactor. It may indicate a secondary V source or a growth of vanadium-bearing particles after release into the atmosphere. By contrast with these

Table VIII. Predicted Vanadium Contributions from Various Sources in the Boston Area

Source	*Element Used as Basis*	*Concn of Element,[a] ng/meter³*	*Predicted V Concn, ng/meter³*	
Continental	Fe	1400	3.6	0.4–3.0
	Mn	30	0.4	
	Al	1430	2.0	
	Sc	0.29	4.1	
	Co	2.4	30	
Coal	Fe	1400	3.5	2–9
	Mn	30	2.3	
	Al	1430	2.6	
	Co	2.4	8.6	
			Venezuelan Crude	*Residual Oil*
Residual oil	Co	2.4	1,400	7,000
	Zn	380	16,300	550,000
	Sb	9	3,650	157,000
	Se	4	1,210	11,600
	Range of Predictions		1200–16,300	700–550,000
Observed Range[a]		90–2400		
Average[a]		600		

[a] Gordon *et al.* (*18*).

curves, those obtained for elements such as Fe and Al, produced largely by continental dust and coal burning, are tilted in the opposite direction, favoring large particles. The size distributions for vanadium-bearing particles found by Rahn (*4*) in remote areas of Canada are generally similar to those of other elements produced by continental weathering, as expected from the calculations of Table IV. This observation does not rule out the possibility that a portion of the vanadium results from residual oil combustion in distant cities, as the size distribution of V-bearing particles may be altered considerably during atmospheric transport over long distances and being subjected to rain-out and wash-out along the way. It would be of considerable interest to determine the size distributions of the particles in areas such as the Pacific Ocean, where our calculations suggest that two-thirds of the vanadium originates from man's activities, principally residual oil burning.

Summary and Conclusions

Newly developed analytical techniques, particularly neutron activation analysis using Ge(Li) γ-ray detectors, have so increased the sensitivity for observation of vanadium in atmospheric particulate material that one can now measure its concentration at any ground-level site on earth, including Antarctica, where values as low as 0.7 pgram/meter^3 (7×10^{-4} ngram/meter^3) are observed. Only about one-third of the 0.35 ngram/meter^3 of vanadium observed in windward Hawaii samples can be accounted for by natural sources, and the remaining two-thirds originates from man's activities. Because of site-to-site variations of observed V concentrations and large uncertainties in predicted V contributions for rural Canada, we are unable to say if some of the vanadium originates from anthropogenic sources. Urban areas in the U. S. can be divided rather clearly into two groups on the basis of their atmospheric V

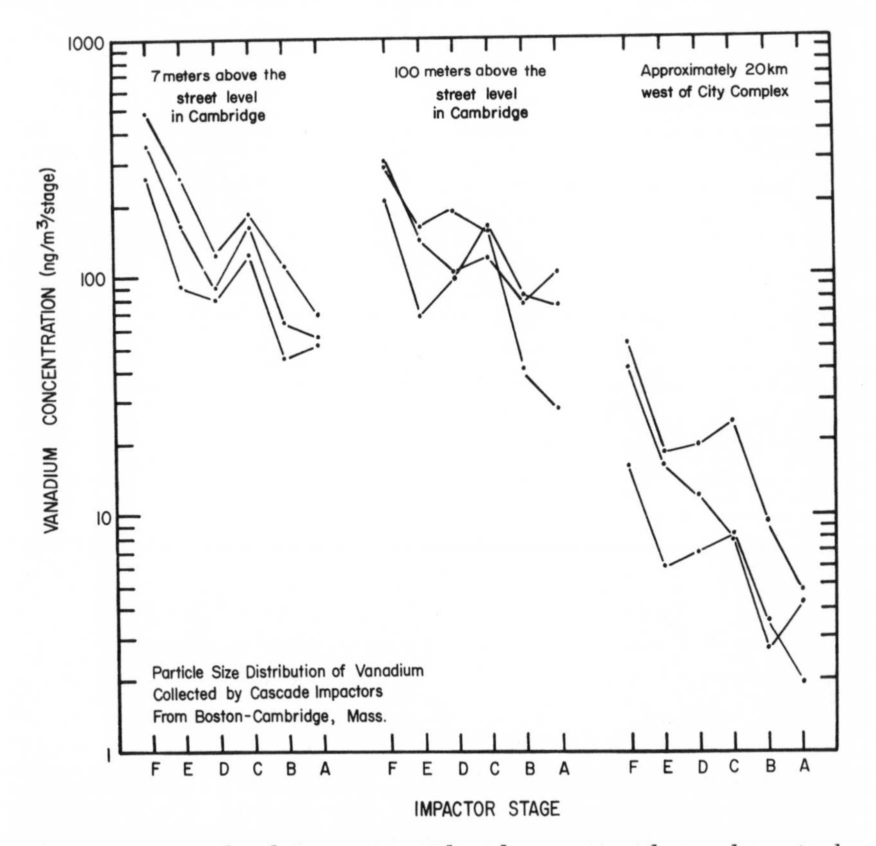

Figure 1. Size distributions of vanadium-bearing particles as determined by analysis of material collected with a six-stage cascade impactor. The weight of vanadium collected per stage per cubic meter of air is plotted vs. *stage designation of the impactor. The largest particles are collected on Stage A and the smallest on Stage F. The first two sites were on the M.I.T. campus in Cambridge, Mass., and the third site was in Wellesley, Mass., about 1 mile west of the circumferential highway, Rt. 128.*

Table IX. Total Amounts of Vanadium Contained in Coal and Oil Burned Annually in the Boston Area (1966)

	Total Consumed[a]	
Use	*Coal, Metric Tons*	*Residual Oil, 10^6 Gallons*
Residential heating	156,000	—
Commercial	59,000	437
Manufacturing	78,000	196
Steam-electric plants	20,000	730
Transportation	29,500	23
Total	342,500	1386
Vanadium concn, ppm	21[b]	870[c]
Total vanadium, metric tons	12	4100

[a] Morgenstern *et al.* (*23*).
[b] Zubovic *et al.* (*24*).
[c] Gladney and Baumgartner (*22*).

concentrations, one group having levels of about 5 to 20 ngrams/meter^3 and the second having concentrations between 150 and 1400 ngram/meter^3. With very few exceptions, the high-vanadium areas occur in the northeastern United States where large amounts of residual oil, generally containing several hundred parts per million of vanadium, are burned for heat and electric-power generation. Concentration patterns of source materials in the Boston area show that residual oil combustion is the only source material with a high enough ratio of V to Fe, Mn, and other elements to account for the high relative observed V concentration in the atmosphere. One potential source of vanadium, the steel industry, is shown to be a relatively unimportant source of atmospheric vanadium.

Murozumi, Chow, and Patterson (*25*) have previously shown that atmospheric lead, largely from combustion of leaded gasolines in internal-combustion engines, has increased on at least a hemispheric scale and can thus be used as a tracer for large-scale atmospheric movement of anthropogenic particulate material. There is also evidence for an increase in the deposition of mercury in Greenland snow strata during the past century or so (*26*), but a definite link with man-made sources has not yet been established. Analysis of vanadium concentrations in the area of the Pacific Ocean strongly suggests that vanadium is an additional metal that can be used as a tracer for hemispheric movement of particulates generated by certain of man's activities, especially in polar regions of the northern and southern hemispheres. Size distributions of vanadium-bearing particles, although difficult to measure in such clean atmospheres, would help to clarify changes in particle sizes during transport over long distances. Although lead is a good indicator of the movement of man-made aerosols, its sources are widely distributed over all areas of the world where there is appreciable motor vehicle traffic. By contrast, the sources of vanadium are more localized in areas where large amounts of residual oil are burned.

The atmospheric concentrations of Zn and Sb are also anomolously high in remote areas. Rahn has suggested that major fractions of these and several other elements (*e.g.*, Se, Hg, Cu) present in the atmosphere at remote sites may originate from man's activities (*4*). If so, these elements may serve as more useful indicators of anthropogenic aerosols than vanadium since their enrichments relative to the world-wide soil average are much greater than that of vanadium. However, before they can become useful tracers it will be necessary to establish clearly their major

Table X. Estimated Annual Rates of Global Injection of Vanadium into the Atmosphere from Petroleum and Natural Sources

Source	*Total Amount Injected or Consumed Annually,[a] 10^6 metric tons*	*Assumed V Concn, ppm*	*Total V Injected, metric tons*
A. Natural			
Soil and rock dust	200	135	2.7×10^4
Volcanic debris	50	200	10^4
Sea salt	80	0.15	12
Total natural			3.7×10^4
B. Petroleum	2000[b]	50	2×10^4[c]

[a] SMIC Report (*28*).
[b] U. S. Bureau of Mines (*29*).
[c] Assumes half of the V in petroleum is released to the atmosphere.

sources, which are presently not known. In several cases (*e.g.*, Hg, Se), the elements themselves or some of their compounds are fairly volatile, so it is possible that their atmospheric enrichments could result from entirely natural processes or subtle activities of man. For example, Weiss, Koide, and Goldberg (*26*) have suggested that recent increases in Hg deposition in Greenland snow may result from its release from soils disturbed by widespread agricultural activity. Thus, careful studies should be made to establish the sources of these elements, as one or more of them could become useful tracers of anthropogenic aerosols, if one had definitive knowledge of their origins.

Bertine and Goldberg (*27*) have shown that the rate of man's mobilization of vanadium (considering only fossil-fuel combustion) is far smaller than that of natural processes as measured by the outflow rate of vanadium in the world's rivers or its sedimentation rate at ocean floors. Their conclusion is probably correct although their calculations do not indicate the relative rates of natural and man-made injection of vanadium into the atmosphere. We have attempted a crude estimation of these rates in Table X. The amounts of soil and rock dust, sea salt, and volcanic debris injected into the global atmosphere annually fall with the wide ranges of values given in the SMIC report (*28*). We have assumed that half of the vanadium present in crude petroleum is ultimately released to the atmosphere. Although this calculation is admittedly crude, it indicates that comparable amounts of atmospheric vanadium arise from petroleum combustion and from natural sources. This supports the results of our analysis of the windward Hawaii data. The relative petroleum contribution to the world-wide atmospheric vanadium concentration may be considerably greater than indicated in Table X because of the small sizes of vanadium-bearing particles from oil combustion. These smaller particles may have greater atmospheric residence times than the larger particles of continental dust, sea salt, and volcanic debris.

Acknowledgments

This work was partly supported by the U. S. Atomic Energy Commission under Contract No. AT-(40-1)-4028 and by the National Science Foundation under Contract No. GV-33335. Computer use was supported in part by National Aeronautics and Space Administration Grant No. NSG-398 to the Computer Science Center of the University of Maryland. Special thanks are given to the staff of the National Bureau of Standards Reactor for the excellent neutron irradiations.

Literature Cited

1. Zoller, W. H., Gordon, G. E., *Anal. Chem.* (1970) **42**, 257.
2. Hoffman, G. L., Ph.D. Thesis, University of Hawaii, 1971.
3. Hoffman, G. L., Duce, R. A., Zoller, W. H., *Environ. Sci. Technol.* (1969) **3**, 1207.
4. Rahn, K., Ph.D. Thesis, University of Michigan; U. S. Atomic Energy Commission Report, **No. COO-1705-9,** May 1971.
5. Mason, B., "Principles of Geochemistry," 3rd ed., Wiley, New York, 1966.
6. Vinogradov, A. P., "The Geochemistry of Rare and Dispersed Elements in Soils," 2nd ed., English translation, Consultants Bureau, New York, 1959.
7. Gladney, E. S., Zoller, W. H., Duce, R. A., Jones, A. G., *Antarctic J. U. S.*, in press.
8. John, W., Kaifer, R., Rahn, K., Wesolowski, J. J., private communication, Aug. 1971.
9. Rahn, K. A., Dams, R., Robbins, J. A., Winchester, J. W., *Atmos. Environ.* (1971) **5**, 413.
10. Harrison, P. R., Rahn, K. A., Dams, R., Robbins, J. A., Winchester, J. W., Brar, S. S., Nelson, D. M., *J. Air Pollut. Control Assoc.* (1971) **21**, 563.

11. Brar, S. S., Nelson, D. M., Kline, J. R., Gustafson, P. F., Karabrocki, E. L., Moore, C. E., Hattori, D. M., *J. Geophys. Res.* (1970) **76**, 2939.
12. Keane, J. R., Fisher, E. M. R., *Atmos. Environ.* (1968) **2**, 603.
13. Pillay, K. K. S., Thomas, C. C., *J. Radioanal. Chem.* (1971) **7**, 107.
14. National Air Pollution Control Administration, "Air Quality Data from the National Air Sampling Networks and Contributing State and Local Networks, 1966 Edition," APTD 68-9, 1968.
15. Morrow, N. L., Brief, R. S., *Environ. Sci. Technol.* (1971) **5**, 786.
16. Kneip, T. J., Eisenbud, M., Strehlow, C. D., Freudenthal, P. C., *J. Air Pollut. Control Assoc.* (1970) **20**, 144.
17. Moyers, J. L., Zoller, W. H., Duce, R. A., Hoffman, G. L., *Environ. Sci. Technol.* (1972) **6**, 68.
18. Gordon, G. E., Zoller, W. H., Gladney, E. S., Jones, A. G., Hopke, P. K., presented at the 163rd National Meeting of the American Chemical Society, Boston, Mass., April 1972.
19. Abernathy, R. F., Peterson, M. H., Gibson, F. H., "Major Ash Constituents in U.S. Coals," *U. S. Bur. Mines Rept. Invest.* (1969) **7240**; "Spectrochemical Analyses of Coal Ash for Trace Elements," *U. S. Bur. Mines Rept. Invest.* (1969) **7281.**
20. Filby, R. H., Haller, W. A., Shah, K. R., *J. Radioanal. Chem.* (1970) **5**, 277; Shah, K. R., Filby, R. H., Haller, W. A., *Ibid.* (1970) **5**, 413.
21. Filby, R. H., Shah, K. R., in "Proceedings of the American Nuclear Society Topical Meeting on Nuclear Methods in Environmental Research," Vogt, J. R., Parkinson, T. K., Carter, R. L., Eds., pp. 86–96, University of Missouri, Columbia, Aug. 1971.
22. Gladney, E. S., Baumgartner, G., University of Maryland, unpublished data, 1972.
23. Morgenstern, P., Goldish, J. C., Davis, R. L., "Air Pollutant Emission Inventory for the Metropolitan (Boston) Air Pollution Control District," prepared by Walden Research Corp. for the Mass. Department of Public Health, June 1970.
24. Zubovic, P., Stadnichenko, T., Sheffey, N. B., *U. S. Geol. Surv. Prof. Pap.* (1960) **400-B.**
25. Murozumi, M., Chow, T. J., Patterson, C. C., *Geochim. Cosmochim. Acta* (1969) **33**, 1247.
26. Weiss, H. V., Koide, M., Goldberg, E. D., *Science* (1971) **174,** 962.
27. Bertine, K. K., Goldberg, E. D., *Science* (1971) **173,** 233.
28. "Report of the Study of Man's Impact on Climate (SMIC)," M.I.T. Press, Cambridge, Mass., 1972.
29. U. S. Bureau of Mines, "Mineral Facts and Problems," *U. S. Bur. Mines Bull.* (1970) **650.**

RECEIVED September 5, 1972.

4

The Three-Phase Equilibrium of Mercury in Nature

EVALDO L. KOTHNY

Air and Industrial Hygiene Laboratory, State of California Department of Public Health, 2151 Berkeley Way, Berkeley, Calif. 94704

Mercury levels are postulated to be the result of equilibrium between the contents in the atmosphere, particulate matter, and rocks. Mercury is released into air by outgassing of soil, transpiration and decay of vegetation, and by heating processes. Most mercury is adsorbed onto atmospheric particulate matter. This is removed from air by dry fallout and rainout. Humic material forms complexes which are adsorbed onto alluvium, and only a small soluble fraction is taken up by biota. Small clay particles and rainout particles are distributed throughout the oceans because of the slow settling speed. Pelagic organisms agglomerate the mercury-bearing clay particles, thus promoting sedimentation and acting as one source for mercury for the midoceanic chain. Another source is the uptake of dissolved mercury by phytoplankton and algae.

This paper develops a model for the dynamics of mercury exchange in nature assuming an equilibrium between mercury in contiguous phases. Exchange of mercury between the different components of the environment (*i.e.*, water, air, soil) is highly complex, and, as a basis for calculation, not only the absolute value of the concentrations in the various components but also the transfer relationships are important. There are many weak points in our knowledge of the parameters; therefore some of the numbers presented are far from precise. The information referenced with the calculations will guide the reader. Material previously reviewed is not included, except for a short guide to the most practical analytical methods for mercury.

In explaining the cycle of mercury in nature, atmospheric mercury is considered first. The natural processes are placed before man-made processes. Therefore, natural evapo-transpiration of mercury from soil and losses from vegetation are considered before the release of mercury by heating, drying, or burning.

Because mercury adsorbs strongly onto surfaces, the equilibrium conditions between atmospheric and particulate matter are analyzed next. This is the most important process for the removal of mercury from the gaseous phase, but the limited information available cannot provide a strong basis for relationships which have been observed. Once dust settles by dry fallout or rainout, mercury goes back to soil or is washed into water bodies.

Discussion on soil and alluvium centers on the organic complexing with humic acid and adsorption of these complexes onto clay. The cycle within the interior of the earth, leading to deposits, is briefly mentioned.

Runoff as well as rainout carries soluble and adsorbed mercury into freshwater bodies and oceans. The very low levels of mercury available from the small solubility of humates and other mercury complexes are the source of mercury for land and water plants. Aquatic organisms derive a part of the mercury by ingestion of water plants, such as phyto-plankton. Another source of mercury for aquatic organisms is zöoplankton, which derives its mercury by ingestion of clay particles. These clay particles are agglomerated and then sedimented as part of the plankton excrement.

Methylating bacteria transform a small portion of mercury in alluvium into alkyl mercurials which, because of lipotropy, have a long half-time and increased toxicity in animals. The diverse mercury compounds differ in toxicity according to the species involved and route of entry. Selenium seems to be a biogeochemically important element which counteracts toxicity of naturally occurring mercury.

Finally, the cycle which summarizes the equilibrium of mercury in nature is presented. The cycle can be considered closed through the physical transfer of mercury in deep sea sediments to the continents by plate tectonics, where mercury will eventually be remobilized by magmatic heat, erosion, and mining. Since some of the geological and chemical terms may be unfamiliar to the reader, a glossary is included.

Analytical Methods and Sources of Information

As a basis for calculation of the mercury cycle, measurements presented in the tables are coupled with the information of excellent reviews (*1, 2, 3, 4, 5, 6*) and observations by S. H. Williston (*7*). Information on the distribution of mercury in nature was very limited before 1930 because adequate microanalytical methods were not available. During the 1930's, Stock (*8*) analyzed the mercury content in many common materials using a microgravimetric method. The use of mercury and its compounds boomed after World War II, and the consequent development of new analytical techniques accounts for the large increase in the amount of data concerning environmental mercury in the following decades (*4, 5*). The concern about mercury as an environmental hazard has recently stimulated the introduction of new and simpler instrumental techniques. These techniques do not give accurate results in all applications. Sample gathering and preparation continue to be sources of error in the new as well as in the older methods of analysis. Spiking is not the answer for demonstrating the recovery of mercury in a particular analytical procedure. Mercury is adsorbed on plastic (*9*), glass (*10*), and paper (*11*); it evaporates from open containers (*12*), it is trapped inefficiently by some sampling devices (*13*), and much of it is lost if samples are dried. Interlaboratory testing on standard reference material has been made, and results agree; however, this testing necessitates standard drying and careful handling procedures. A different situation exists when real samples are analyzed. For example, mercury in soil may exist in a strongly complexed form (*14*) or may even be in a more volatile form, which is lost upon drying (*15*). Consequently, results obtained by methods using a minimum of sample preparation such as neutron activation analysis, have been higher than those obtained by chemical methods. Conversely, chemical methods may give systematically higher results because of accidental introduction of mercury with the reagents (*11*). However, when mercury in the sample is completely transferred to a solution as the mercuric ion (*12, 13, 16, 17*), any sensitive detection technique will work, including colorimetry and flameless atomic absorption (*4, 5, 17, 18*).

Some alkyl mercurials have been analyzed by gas chromatography, and several modifications have been proposed to minimize analysis time (*4, 5*). The fastest separation, which is readily applied to large sets of samples, has been accomplished by diffusion to a cysteine-impregnated filter paper in a Conway cell (*19*). In another technique, inorganic mercury compounds can be transformed to metallic vapor by adding alkaline stannite to an homogenized sample without mineralization. Organic mercurials are similarly transformed when alkaline stannite and cadmium chloride are added to the substrate. This technique obviates the long separation steps used in gas chromatographic methods. The released metallic vapor is then measured by flameless atomic absorption (*20, 21, 22*).

Published concentration data on plants and tissues have been presented either on a wet or a dry sample weight basis. When this condition is not specified, the data must be viewed with caution. As a result, some of the numbers used here are accurate only to within one decade. On the other hand, the established concentrations of mercury in soils, rocks, and sediments are more accurate.

Disturbance of the Cycle on a Global Scale

The total amount of mercury on the continents to a depth of 1000 meters can be calculated to be 4×10^{10} tons. The main sources of environmental problems from mercury stem from exploitable deposits for which a mean value of 3×10^7 tons is estimated here. This value is 50% larger than that assigned by Saukov (*113*) to accommodate newly discovered deposits. This represents less than 0.1% of the total. The dissolved and suspended mercury content of the oceans is estimated to be in the range of 5 to 20×10^7 tons (*1, 23*). At the present rate of mining (10^4 tons per year) and assuming half of the metal is recycled to the ocean and would stay in solution indefinitely, the mercury content of the oceans would be doubled in 10^4 years. Thus, man may add to the oceans, each year, an amount of mercury equal to less than 0.01% of their current level. The volatility of mercury causes a significant fraction to enter the atmosphere. Background levels range from 0.7 to 15 ng/m^3 (these values are extremes for oceanic air and soil air observed by Williston (*7*)). Assuming a global mean of 2 ng/m^3, the total atmospheric loading would be 10^4 tons (2 ng/m^3 $\times$ 6000 $\times$ 10^6 km^3 $\times$ 10^9 m^3/km^3 $\times$ 10^{-15} ton/ng).

Burning of fuels and heating of materials which contain small amounts of the metal create man-made emissions of about 10^4 tons/year. Thus, the background concentration in air would about double the first year and reach alarming concentrations (50 μg/m^3) in about 10^4 years at present emission rates if no removal processes existed. Fortunately, there is a continuous interchange between mercury in the atmosphere, soil, and water. The equilibrium between these three phases is reviewed below in relationship to biota to assess the actual accumulation of mercury in atmospheric and aqueous environments.

Mercury in the Gas Phase

Atmosphere. From any soluble or insoluble mercury compound a measurable amount of mercury evaporates. The volatility of the compounds is strongly influenced by their solubility in water, adsorption on surfaces, and the relative humidity of the air. For example, 50 times more mercury is volatilized from the sulfide at 24°C at a relative humidity of air of 100% than at 0% (*24*). The relative amounts volatilized from other compounds are given in Tables I and II. The alkyl and aryl mercurials are by far the most volatile.

Table I. Volatility of Inorganic Mercury Compounds[a] at 24°C

Compound	*Atmospheric Concentration of Mercury, μg/m³*
Sulfide in dry air (RH below 1%)	0.10
Sulfide in damp air (RH close to saturation)	5.0
Oxide in dry air (RH below 1%)	2.0
Iodide in dry air (id.)	150
Iodide coprecipitated on cuprous iodide (10% HgI_2 + 90% CuI), in dry air (id.)	$\leqslant 5$
Iodide—crystal violet complex (HgI_2.CV) in dry air (id.)	0.4
Calcined pyrophosphate (heated to 700°C), in dry air (id.)	7
Calcined pyroarsenate (heated to 700°C), in dry air (id.)	6
Cyanide (RH approximately 30%)	15
Fluoride (RH approximately 30%)	20
Fluoride (RH below 1%)	8

[a] Author's data with method from Ref. *17* obtained by pulling air at 2 liters/min through 100 ml of 4-mm glass beads coated with the substance indicated.

Because of the volatility of the mercury compounds, mercury is constantly released over the continents by transpiration from soils and land plants. On a global scale, soils can be estimated at 93% of the total land surface. Land plants represent about 66% of the total biomass production on this planet (*25*). Transpiration from vegetation is readily observed in the first few minutes following sunrise when the stomata open and release the mercury which accumulated during the night (*7*).

The combined effect of transpiration from soil and vegetation can be illustrated by the following oversimplified example. With strong westerly winds (25 to 40 km/hr), the concentration of mercury in oceanic air increases from 0.7 to 2 ng/m³ near California's Central Valley (*7*), a distance of about 100 km inland from the ocean. With this number the release of mercury per km² per day can be calculated. For convenience consider a land section 0.01 km wide with an inversion height of 0.4 km. In this example the distance from the coast to the valley is filled with oceanic air in about 3 hours. The surface which liberates the mercury is thus 1 km², and the corresponding air volume is 0.4 km³. Because this volume is swept in 3 hours, the volume swept by oceanic air in 24 hours is 8 times larger or 3.2 km³. The mercury increased by 1.3 ng/m³. The amount of mercury released over the continent was thus estimated to be:

$$3.2\ \text{km}^3 \times 10^9\ \text{m}^3/\text{km}^3 \times 1.3\ \text{ng/m}^3 \times 10^{-15}\ \text{ton/ng} =$$
$$4 \times 10^{-6}\ \text{ton/day-km}^2$$

This value cannot be extrapolated to all regions of the earth because soils in California are about five times higher in mercury (Table V, Franciscan sediments) compared with soils in other parts of the world. Climate is another factor which may influence this value. Therefore, we may assume that about one-fifth of that amount is released as a mean over the continental soil surface. Thus, the global value on a per day basis would be:

$$148 \times 10^6 \text{ km}^2 \times 4 \times 10^{-6} \text{ ton/day-km}^2 \times 1/5 = 120 \text{ tons/day}$$

A similar value (100 tons/day) has been calculated using an iterative process on evaporation and transpiration from individual continents and considering climatic factors. This value has been inserted in the cycle (Figure 1).

It is not known what amounts of gaseous metallic mercury are contributed by biotransformation and evaporation from the sea. We may assume that the lowest concentration (0.6 ng/m^3) observed by Williston (*7*) with an airborne instrument over the ocean is an equilibrium value. With an onshore breeze greater values up to 1.5 ng/m^3 observed near shore may have been formed by photolysis (*3*) of alkyl mercurials. The

Table II. Volatility of Organic Compounds[a] at 22–°23°C

Compound	*Conditions*	*Atmospheric Concentration of Mercury, μg/m³*
Methylmercuric chloride, aqueous	0.08% in 0.1*M* phosphate buffer, pH 5	900
	0.04% in 0.1*M* phosphate buffer, pH 5	230
	0.04% in 0.5*M* phosphate buffer, pH 5	560
	0.04% in 0.1*M* phosphate buffer, pH 7	140
Methylmercuric dicyandiamide, aqueous	0.04% in 0.1*M* phosphate buffer, pH 5	140
Phenylmercuric acetate	solid, *ca.* 30% RH in air	140
	solid, dry air with < 1% RH	22
Phenylmercuric nitrate	solid, *ca.* 30% RH in air	27
	solid, dry air with < 1% RH	4
Cysteine mercuric complex, acid form	solid, air saturated with water vapor	13
	solid, dry air with < 1% RH	2

[a] Author's data with method from Ref. *17* obtained by pulling air at 2 liters/min through solutions or 100 ml of 4-mm glass beads coated with the dry substance indicated.

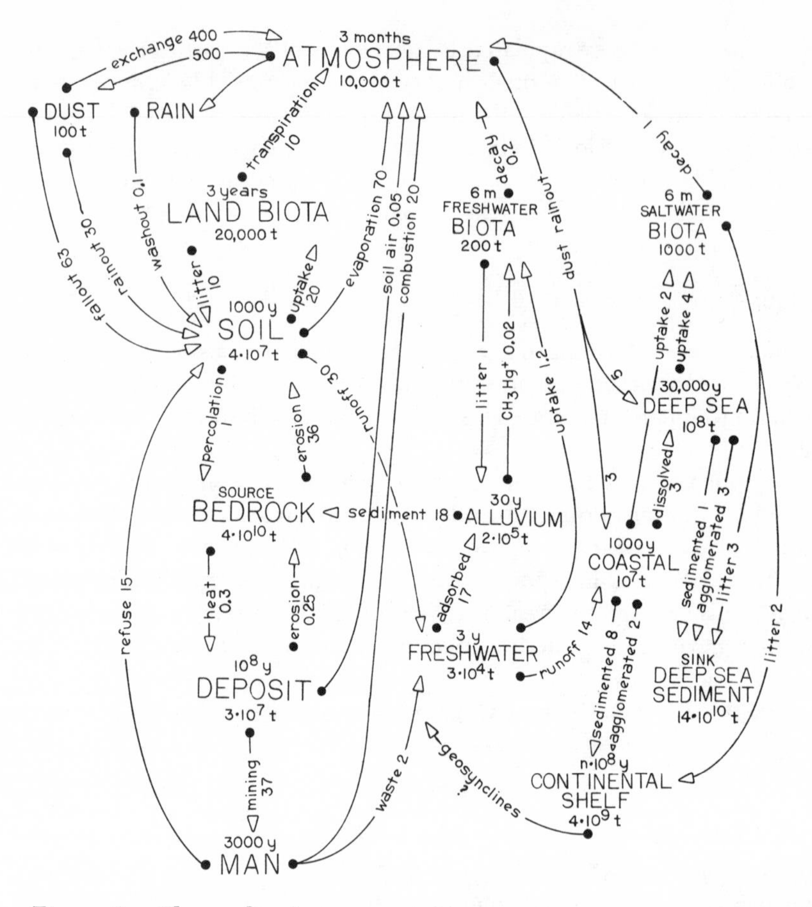

Figure 1. The cycle of mercury in nature. Dots represent sources and arrowpoints are the sink for each route. Numbers on the routes are tons per day. Residence time in years (y) and months (m) are above each entity. Immobilized tonnage (metric, t) is indicated below each entity.

source of alkyl mercurials near oceanic shores may be decaying saltwater biota and photosynthetic activity in algae. The amounts released from saltwater biota are small compared with the amount lost to sedimentation of detritus. Instruments used for these observations were sensitive to metallic mercury vapor only.

Atmospheric mercury in urban areas results from various sources. Stack emissions in California's San Francisco Bay area may have caused the short term peaks of 30 to 50 ng/m³ observed when the background ranged from 3 to 8 ng/m³ on days with low and variable winds (*7, 26, 27*). Most atmospheric mercury in urban areas (Table III, urban) is contributed by processes involving heat which vaporize mercury contained in fuel and other materials (*28*). Noteworthy are the contributions from steel plants (*29*), ore smelting (*3*), power plants (*30, 31*), and ceramic and cement manufacture and space heating (*6*). All fuels contain significant amounts of the metal (*1, 3, 30, 32*). Background levels (Table III, rural) are contributed by natural sources such as transpiration from vegetation, drying up of soil, natural steam wells, and forest fires. Large scale agricultural operations contribute a small amount because of the use of sprays containing mercury compounds and slow release of residual mercurials after application.

The level of mercury in ambient air inside homes and offices is much higher than that of urban outside air (*33*) (Table III, indoors). This may be partly the result of the extensive use of poly(vinyl chloride) and other synthetic resins, which, in some cases, are made with a mercury catalyst whose residues might volatilize. Paper fabricated with pulp containing a mercurial fungicide also contains that metal. Latex paint is protected from mold by addition of a mercurial. Slow evaporation of transformed residuals in paints can be blamed for most indoor gaseous mercury (*33, 34*). Annually, several tons of mercury are also lost by thermometer breakage (*35*). Mold-free cement, used for grouting bathroom tiles, also contains substantial amounts of a mercury fungicide.

The limit of 50 $\mu g/m^3$ adopted by the American Conference of Governmental Industrial Hygienists (*36*) as the upper safe level for industrial exposure (Table III, industrial and commercial) is maintained within limits by chloralkali plants, dentists, and manufacturing operations. Exposure to higher levels still occurs in some mining enterprises, university laboratories, and elsewhere. Chloralkali plants using mercury cells to produce Na-amalgam, which later is decomposed to sodium hydroxide and hydrogen, have been regarded as the largest sources of mercury in liquid discharge as well as atmospheric output. Much of the mercury in the effluents (30 to 40 grams mercury were lost for every ton of chlorine produced) has actually been reduced to a tolerable level (currently 0.5 gram mercury is lost for each ton of chlorine produced). Reductions in the content of the mercury-saturated hydrogen produced in the decomposers and ventilation from the cells have been larger than tenfold (—*e.g.*, the loss of mercury in the hydrogen vented to waste was reduced from a previous maximum of 10 grams to 0.4 gram per ton of chlorine produced). More reductions are feasible in the near future (*2*).

The natural dispersion mechanism of mercury as explained above overshadows input into the global environment by human activity, the latter being of prime importance in localized environments (*37*), generally encompassing centers of high population density. In meteorological terms the horizontal dispersion is several orders of magnitude greater than the vertical (*38*), and therefore most fallout will occur near the place of emission (*39*). Nevertheless, evidence of mercury pollution has

Table III. Sources of

Sources

Rural
- Vegetation (*7, 17, 19, 75*)
- Soil (*1*)
- Seed dressing and sprays, soil levels increase
- Volcanic activity, fumaroles, and steam wells (*110, 111*), maximum limits

Urban
- Fossil fuels (*1, 30, 31*)
- Firewood
- Powerplants (1 ton/year each (*30*))
- Incinerators and sewage plants
- Iron ore processing (*29*)
- Ceramic manufacture
- Ore smelting

Indoor
- Paint (*27*) and grout (as fungicides)
- Plastic, catalyst residues in PVC tiles
- Kraft paper (filter paper (*11*))
- Spillings (from broken thermometers (*33*))

Industrial and commercial
- Chloralkali plants (*31*); loss of mercury for each ton of chlorine (cell ventilation, hydrogen vent)
- Dental offices (*33, 112*)
- Manufacture of batteries, switches, chemicals, etc.

been found in ice deposits in Greenland (*40*). The values reported represent the equilibrium between mercury in the atmosphere, dust, and ice (Table VII). It is not known, however, if most of this metal originated from the heavily industrialized areas on the Northeast Atlantic Coast or from elsewhere.

Gas-to-Solid Phase Transfer

Dust. Williston (*26*) mentioned the possibility of mercury removal by dust particles. He based his assumption on the observation that an open silica gel container adsorbed up to 1 μg/gram from laboratory air after several months of exposure (*7*). In laboratory air, half of the total mercury was found on suspended particulate matter when the latter was collected on a membrane filter at 40–60% relative humidity (*24*). The settled dust had a lower mercury content (Table IV). Thus, mercury on suspended particulate matter appears to be associated mainly with the submicron particles. Other examples in Table IV show higher levels of mercury collected on urban dust. Apparently some samples from

Atmospheric Mercury Vapor

Source Content (Approximate)	*Associated Ambient Concentration, ng/m³*
	0.7 to 9 (*7*)
0.00 to 1 μg/g	
0.01 to 2 μg/g	
0.07 to 0.25 μg/g[a]	
4 and 28 μg/m^3	
	2 to 60 (*7*, *27*)
0.01 to 20 μg/g	
0.03 to 1 μg/g[c]	
Plumes: 4000 ng/m^3 (*110*)	
Sludge: 5 to 30 μg/g[c]	
Ore: 0.1 to 100 μg/g[c]	
Clay: 0.05 to 2 μg/g[c]	
Ores: 0.1 to 100 μg/g[a] (*113*)	
	30 to 300[a, b] (*33*)
10 to 500 μg/g[c]	
0.35 to 1.2 μg/g[a]	
1 to 20 μg/g[c]	
several tons/year (*35*)	
	up to 50,000[b] (*36*)
0.2 gram/ton	

[a] Author's data by method in Ref. *17*.
[b] Measured indoors.
[c] Author's estimate.

urban areas were collected when the sampler was shrouded in smoke (*i.e.*, New York City) or exposed to a slow moving plume (*i.e.*, Cincinnati and Charleston). Unfortunately we do not know the location and the meteorological conditions during that sampling. Therefore, these special examples should be regarded as exceptions. It can be demonstrated that these values are easily obtainable by assuming fuel containing 0.5 μg/gram is burned with a 100% excess air. The exhaust will contain approximately 10% CO_2, which corresponds to nearly 0.5 kg fuel per m^3 and to 250 μg of mercury. If the plume dilutes 250 times before reaching the sampler, the CO_2 will have dropped to 0.1% and the mercury to 1000 ng/m^3. Although no precise information can be extracted from the numbers in Table IV, it is safe to infer that the equilibrium ratio between particulate matter and gaseous mercury may reach 1:10. If such is the case, the content in dust will be equivalent to 100 ng/m^3, a value similar to the one recorded for Cincinnati. With less dilution of the plume, higher values similar to those observed in New York City homes or outdoors downwind from a major source could also be justified.

Table IV. Mercury in Particulate Matter

Location	*Gaseous Mercury Concentration, ng/m³*	*Particulate Mercury Concentration* μg/g	*Particulate Mercury Concentration* ng/m³
Indoors			
Window sill [a]	170	100	[c]
Home in N.Y.C. (*77*)	[c]	[c]	1000 to 41,000
Laboratory (suspended) [a]	30	800	30
Laboratory (ceiling, by impaction) [a]	30	50	[c]
Laboratory (floor) [a]	30	6	[c]
Outdoors			
San Francisco Bay, dry weather [a]	6	0.3	[c]
San Francisco Bay, foggy [a]	6	6	1 [b]
Chicago (*41*) foggy	[c]	110	8 [b]
Chicago (*114*) clear	[c]	10	1 [b]
N.Y.C. (*77*)	[c]	[c]	1000–14,000
Cincinnati (*77*)	[c]	[c]	100
Charleston (*77*)	[c]	[c]	170

[a] Author's data by method in Ref. *17*.
[b] Author's estimate.
[c] Not determined.

Other possible relationships can be extracted from Table IV. For instance, the mercury content of particulate matter obtained by filtration in metropolitan areas appears to be maximized when sampling under humid and reducing atmospheric conditions (*7, 41, 42*) as opposed to the samples collected under dry and oxidizing conditions. This last point can be clarified when one considers that mercury in either the gas or solution phase is readily adsorbed on surfaces (*6, 24*). This route of transfer decreases with increasing acidity (*43*), decreases with the oxidation potential of the solution (*10*) or gas phase, and increases with the humidity. In this context, mercury shows a striking similarity to organic pesticides (*44*). More information on this interesting removal effect under natural and metropolitan conditions is desirable.

Gas-to-Liquid Phase Transfer

Rain. Mercury in the gas phase is less readily taken up by water. In deaerated water under nitrogen the solubility is 20 to 30 μg/liter (*7*). In distilled water saturated with air at 27°C it is 47 μg/liter (*24*). The washout efficiency of rain can be calculated from the ratio of the pressure of mercury vapor in air to the vapor pressure over liquid mercury at 20°C, which is about 0.2×10^{-6} for an atmospheric concentration of

6 ng/m³. For calculations, a 50-mm precipitation from a cloud 100 m over the surface will be assumed. The 50 liters of rain which fall on each square meter will collect:

$$50 \text{ liters} \times 47\ \mu g/\text{liter} \times 0.2 \times 10^{-6} = 0.5 \times 10^{-3}\ \mu g \text{ mercury}$$

The total contained in the 100 m³ of air will be 600×10^{-3} μg mercury. Therefore, the washout efficiency in this example is only 0.08%. However, assuming the same conditions but a particulate matter concentration of 10 to 50 μg/m³, which collected about 1 ng mercury per m³ (Table IV), calculations show that 16% of the mercury is removed by particle rainout whereas most still remains in the gaseous form. Accordingly, Williston (*7*) did not observe a noticeable reduction of gaseous mercury levels during rainstorms. After rainstorms, however, the mercury level decreased apparently because of the increased ventilation (*7*). The reduction of the mercury level also seems to be related to meteorological factors—*i.e.*, whether a warm or a cold front produced the rainstorm. During cold fronts, ventilation is greater, and the mercury level may seem to decrease during the rainstorm.

Soil

Mercury in soils can be postulated to be the end result of equilibrium between the mercury in the atmosphere, particulate matter, water, and rocks. However, there are exceptions in some parts of the world where final equilibrium has not yet been attained. Indeed, the concentration of mercury in natural soils (93% of all land surface) is anywhere from less than 10 μg/kg to more than 1%. Table V shows a selection of rocks, sediments, and soils from different origins and places to illustrate the most common values associated with the formation of natural soils. Values obtained by analysis of non-soil specimens from California do not differ appreciably from values of similar specimens (*e.g.*, Graywacke and associated sediments) from other parts of the world. Well aerated agricultural soil in plains, undisturbed from metropolitan, industrial, and road emissions and which had never had any spray application is an example of equilibrated soil. Such a soil contains 0.07 μg/gram mercury (*32*). Equilibrium levels of mercury in soils can vary, however. For instance, humic soils tend to carry appreciably more mercury than soils that are low in organic matter (*6, 45*) possibly because of concentration effect arising from litter decomposition. Sediments laid down under strong reducing conditions (negative redox potential or *Eh*) such as coal, peat, shales, petroleum, etc. may contain up to 10 μg/gram (*1, 24*) or more (*28, 46*).

Uplifted oceanic sediments, such as the Franciscan sediments, bear higher amounts than continental sediments whose origin is generally

Table V. Mercury in Soils[a]

Material	*Content, μg Hg/g*
Uplifted oceanic sediments (*1, 47*)	0.1 to 1
Franciscan sediments, Calif.[b]	0.2 to 0.5
Ultrabasic outcrops, Calif.[b]	0.02 to 0.1
Alluvial sediments, Calif.[b]	0.02 to 0.25
Weathered lava, Berkeley Hills, Calif.[b]	0.08 to 0.22
Other igneous rocks, Calif.[b]	0.03 to 1
Soils derived from igneous rocks (*47*), mean	0.07
Soils derived from continental sediments (*32*)	0.07
Coal, shale, petroleum (*1, 24*)	0.01 to 10
Quartz monzonite, Sierra Nevada, Calif.[b]	0.17
Lenses of biotite, derived from above[b]	0.8

[a] On a dry basis.
[b] Author's data by method from Ref. *17*. All values reported are from areas not related to mineralization.

alluvium. Cinnabar deposits, which were formed from hot alkaline sulfide solutions, have commercial value when the content is above 500 μg/gram. Intermediate values from 1 to 500 μg/gram are also known and were formed by impregnations of cap rocks in geothermal fields (*14*) or by erosion haloes. The areas covered by such high concentrations are insignificant when compared with the earth as a whole.

Natural soil derived from weathered igneous rocks contains a mean of 0.07 μg/gram mercury (*47*). Higher concentrations in igneous rocks and soils derived therefrom can be explained by a local equilibrium (*39*). For example, the lavas from the Berkeley Hills, near Berkeley, Calif. (Table V), which contain 0.08 to 0.2 μg/gram may have equilibrated with ultramafic rocks containing 0.02 to 0.1 μg/gram and with Franciscan sediments containing 0.2 to 0.5 μg/gram occurring nearby.

The last two lines in Table V give the mercury content of biotite lenses formed by differentiation in a granite wash and the granite parent material. Elsewhere (*48, 49*) it was also noted that most mercury in granite is contained in the biotite fraction.

It has been repeatedly observed that dilute solutions of mercury are not stable unless the oxidation potential and the acidity of the solution is raised (*6, 10, 43*). For example, equilibration of about 10 grams of Al and Fe hydroxides with a 0.1% solution of mercuric acetate, then rinsing with 10% acetic acid until mercury is absent from the washings, and drying at 50°C for 4 hours, yields Al and Fe hydroxides retaining about 25 μg/gram mercury (*24*).

Sulfur-containing amino acids and proteins form very strong soluble complexes with mercury (*12, 50, 51*). Humic acids are formed by partial decay of organic matter and may have sulfur-bearing residual proteins bound to quinoid polymers, thus acting as strong complexers of relatively low solubility (*6, 52*). The solubility tends to be lowered because of adsorption of humic acids onto clay particles. Adsorption strength depends principally on soil pH. The chemical or physical form of the occurrence of mercury in natural soil can be determined by selective solvent extraction (*6*). None of the most common and stable inorganic and organic substances including the sulfide and organic mercurials were found. It was concluded that the mercury in soil is adsorbed, possibly in the form of a humate as found for coal (*53*). The mercury could be dissolved without destroying organic matter by dilute acidified bromine water or alkaline sulfide.

Alluvium and Internal Waterways. Runoff and spring waters (*1, 46, 54*) are in dynamic equilibrium with humates in the soil. On several well-aerated and drained soils of pH 6 to 7 the ratio for mercury in runoff water over mercury in A_1 horizon soil was equal to 6×10^{-4} (*24*). Therefore, assuming that this value is applicable to any soil, water running off from a soil containing 0.07 μg/gram would contain about 0.04 μg/liter, a value close to that observed on several uncontaminated rivers (Table VIII).

The equilibrium in bodies of water is maintained because mercury is adsorbed onto alluvium. In a typical example, 10 grams alluvium retained 95% of the mercury from natural water containing 0.5 μg/liter after a few minutes contact, the small particles being the most active adsorbents. The most abundant size, accounting for the bulk of the surface area, was the 1- to 12-μ fraction (*43*). In another example, a spring draining a mercury deposit increased the concentration of the head waters of a stream to 136 μg/liter while 50 km downstream the concentration dropped to 0.04 μg/liter (*55*). Similar levels and reductions were observed on rivers with industrial outfall (*1, 56*).

Available information (*23, 57*) points to an annual river discharge of mercury into the ocean of 2500 to 3000 tons. Based on information from "Mercury in Waters of the United States 1970–1971" (*58*) calculations show that these values might be low. From all the rain which falls on the continents, the amount which evaporates (7×10^{13} m^3) concentrates its mercury content (assumed to be 0.08 μg/liter), and this excess concentration is expected to equilibrate with alluvium (3×10^{10} tons). As a result, some of the mercury would remain on land, and only a small fraction would end up in discharge to the ocean. Thus, alluvium containing a background level of 0.07 μg/gram would theoretically end up with:

$$0.07 \text{ gram/ton} + \frac{7 \times 10^{13} \text{ m}^3 \times 0.08 \text{ mg/m}^3 \times 10^{-3} \text{ g/mg}}{3 \times 10^{10} \text{ ton}} =$$

$$0.26 \text{ gram/ton } (= \mu\text{g/g})$$

Thus, mercury transported by alluvium would be about 7800 tons. The dissolved amount in runoff (0.37×10^{14} m^3, containing a world mean calculated at 0.10 μg/liter) would be equal to 3700 tons of mercury. The sum, 11,500 tons of mercury transported annually by water, is about 30 tons/day. On the journey from the source to the estuary, some of the mercury in water exchanges with alluvium, and some alluvium enriched in mercury is diluted with soil. The assumption was made that about 3600 tons of mercury in alluvium (0.12 gram/ton $\times 3 \times 10^{10}$ ton alluvium $\times 10^{-6}$ ton/gram) and 1600 tons dissolved mercury (4×10^{13} m^3 runoff $\times$ 0.04 mg/m^3 $\times 10^9$ tons/mg) enter the coastal waters. Thus, a total of 5200 tons/year = 14 tons/day would go to the ocean (*59*). The ratio of 3600 tons suspended per 1600 ton dissolved mercury, equal to about 2.3, conforms to the mean ratio found in waters of the United States (*58*).

Fine eolian dust carried over the ocean and precipitated by rain (*60, 61*) may also adsorb mercury from the atmosphere. Pelagic organisms may account for the agglomeration of the fine suspended matter in the ocean (*3, 62*), and thus the mercury generated over continents can be transferred to the midocean (*63*). Dissolved mercury in the oceans can be concentrated by phytoplankton (*64*), thus offering another route for the removal and sedimentation of mercury in the midocean. As a result of this removal oceanic surface waters are depleted in this element (*3, 65, 66*) relative to deeper strata, and sediments are accordingly enriched.

Sediments near the coast are lower in concentration because of the larger input of clastic material from the continents (runoff and coastal erosion), but because of a faster turnover (10 to 100 times faster) the sediments near the continental shelves remove a larger tonnage of mercury than pelagic sediments farther out.

Observing production, inventory, and reclaiming numbers (*27, 67*) one can say that possibly half of the mercury mined per year is eventually put back into smaller environments. The release of 6 tons/year into internal waterways by 50 chemical companies, excluding chloralkali plants, does more harm locally (*30, 37*) than a multiple of that amount released into the atmosphere by burning. To complete the picture we may add to this figure the amounts lost by chloralkali plants (around 600 tons/year) (*3*), plus amounts washed by rain, weathering of paint chemicals, seepage of fungicides, percolation from refuse, sewage effluents, etc. Public concern has curbed many of the effluents, but the effect of mercury already adsorbed on alluvium in river systems may be seen for several years on a decreasing scale.

Immobilization of Mercury and Geological Formation of Deposits. After mercury is adsorbed on clay (alluvium) (*43*), it is immobilized on bottom mud under anaerobic conditions (*6, 68*). Acid pH favors biotransformation into methylmercury compounds, but the amount lost is very small (*69*) and of little significance. Moreover, alkyl mercurials are volatile and thermodynamically unstable under natural conditions (*1*) and rapidly photolyze into metallic mercury and free radicals (*3*), returning to the cycle.

Magmatic heat in the presence of water generates alkaline fluids by hydrolysis of silicates. Wherever alkaline solutions interact with reduzates, such as consolidated bottom muds containing sulfides which formed by action of reducing bacteria onto sulfates, the alkaline sulfidic solutions so formed are able to change any mercury compound into highly dispersed sulfide and then to dissolve it slowly from the sediments (*70*). These fluids contain generally 0.1 to 20 μg mercury per liter (*1*). The solubility of mercury in alkaline sulfidic solutions can be represented as being governed by the following reaction:

$$SH^- + OH^- \rightleftarrows S^{2-} + H_2O$$

$$S^{2-} + HgS \rightleftarrows HgS_2^{2-} \text{ (soluble)}$$

Dickson has observed (*71*) that the saturation concentration of mercuric sulfide in the alkaline sulfidic solutions depends on the concentration of sulfide ion (S^{2-}). The concentration of the sulfide ion is a function of pH and the relative excess of sulfhydryl ion (pSH) or free hydrogen sulfide. Some natural fluids contain a pSH several units higher than pOH. Therefore, for practical reasons, the solubility of mercuric sulfide can be expressed as a function of pH, as follows (*24*):

$$\text{Solubility at 25°C (log of mg/liter HgS)} = 1.37\ \text{pH} - 14.3$$

Because the pH shifts to lower values at higher temperatures, the solubility of mercuric sulfide increases when such waters cool down. The deposition of cinnabar takes place, however, by mild oxidation and partial neutralization with carbon dioxide presumably delivered by telluric water (*70, 71*). The emplacement of cinnabar and other associated sulfides depend on the porosity of the rock where such deposits are being formed, the temperature and thermal gradient of the geothermal field, and the composition and extension of the heated rocks (*14*).

Oxidation of the deposits by exposure or erosion generates both gaseous and ionic mercury. Whereas inorganic forms of mercury travel short distances after oxidation (clay or hydroxides may contain 1 to 50 μg/gram), the gaseous mercury is pumped through fissures by the action of barometric pressure changes, reaching the soil surface and forming extensive haloes (*72*). When organic or gaseous mercury enters into contact with hydrocarbons or other hydrophobic organic compounds, a definite enrichment by solvent partition occurs (*1*), as in the earlier example of forest litter (*6, 45, 73*).

In industrial operations, dispersed sulfur, alkaline sulfide, or hypochlorite solutions are effective in trapping mercury. Sulfhydryl resins might be useful for removing alkylated forms of mercury (*74*). The mercury could be regenerated from these traps, or in the case of the dispersed mercury sulfide in sulfur, it might be disposed of in this very immobile form.

The Biological Loop

Land Plants. Observations about low level mercury uptake by plants is a relatively unexplored field with few reports on hand. The information is confounded by many analyses which were performed on oven-dried material. Losses during drying might reduce the fresh weight concentration on some samples by up to one decade. Most plants take up a small amount of ionic or complexed mercury, but gaseous mercury and alkyl mercury compounds appear capable of entering rooted plants much easier (*68*). Some mercury is lost by evaporation through the leaves (*7*). Plants listed as mercurotropic often depend on other trace elements associated with mercury mineralizations for optimal growth. Species of carrots (*19, 75*), allium (*29*), plankton (*64, 66*), algae (*65, 68*), pinus and deciduous trees (*15*), seeds (*75*), as well as roots (*76*) have been reported as effective concentrators of dissolved and gaseous mercury. Uptake of gaseous mercury through leaves (*77*) as well as by interception of dry fallout also takes place (*19*). The extent of uptake appears to depend on soil type, location, and depth. The mercury content of plants growing on normally aerated soil will be different from that from plants collected near decaying metallic sulfides or from those growing in soils with low redox potential.

Most soils may be classified as being aerated. Mercury is contained therein, adsorbed on clay particles or on humic material, in ready contact with atmospheric oxygen. Examples in Table VI show that plants growing on aerated soils take up an amount which varies from less than 0.1 to 0.7 μg/gram on a fresh weight basis, regardless of the concentration of the soil. The first three examples show a common weed which grew on the countryside, on a parking lot in the city, and on a road near a mercury mining town, respectively. The second sample may have been affected by dry fallout. All our samples were analyzed directly without previous drying, and the concentration is referred to a fresh weight basis.

Soils found near sulfidic base metal deposits under active decay may have three principal kinds of mercury compounds: (1) organically complexed and adsorbed on clay, as in aerated soils, (2) inorganic salts,

Table VI. Mercury in Plants

Type of Soil	*Soil Content, μg Hg/g*	*Plant Species and Conditions*	*Plant Content μg Hg/g*
Aerated	0.02	*Melilotus i.*[a], [d]	0.2
	0.15		0.7
	53		0.2
	0.20	*Papaver e.*[a], [d]	0.25
	0.7		0.14
	0.14	cupressus[a], [d]	0.18
	0.04 to 0.65	cedar[c] (*1*)	0.5
	[e]	wheat[c] (*3*)	0.08–0.25
	[e]	oats, leaves[c] (*76*)	0.18
	[e]	oats, seeds[c] (*76*)	0.01
	[e]	peas, seeds[c] (*76*)	0.00
	[e]	radish, tops[c] (*76*)	0.24
	[e]	kale[a] (*115*)	0.15
	[e]	orchard leaves[c] (*116*)	0.16
	[e]	roses leaves[c] (*77*)	0.07
	[e]	roses petals[c] (*77*)	0.20
	[e]	tobacco leaves[c] (*117*)	0.19
Near sulfides under decay	100 to 10,000	deciduous trees [b], [c]	10
	10 to 100		3
	0.1 to 0.3	coniferae[b], [c]	0.2–0.8
	1 to 40	(*1*)[c]	0.5–3.5
	0.5	cupressus[a], [d]	0.9
Near H_2S bearing wells (Wilbur Springs, Calif.)	0.19	*Papaver e.*[a], [d]	0.04
	0.2 to 100	deciduous trees[a], [d]	0.01–0.04
Aquatic (ocean at Santa Monica, Calif.)	[e]	mixed algae[a], [c]	0.12
Ocean	0.15 μg/l	mixed algae[c] (*3*)	0.023–0.037

[a] Author's data by method from Ref. *17*.
[b] Author's mean values calculated from Ref. *15*.
[c] Analysis on previous dried samples and referred to dry weight.
[d] Analysis on fresh samples and referred to the wet weight.
[e] Not determined.

generally hydrolytic decomposition products of the chloride, and (3) gaseous metallic. Whereas plants possess a barrier to the uptake and circulation of the first two kinds of compounds because of strong adsorption on clay, humic material, and other biologic substances within the plant, available information indicates that apparently many plants have no barrier against uptake of gaseous mercury through the roots. Therefore, these plants contain high levels of the metal. This uptake may even lead to metallic droplet formation in woody parts and seeds (*1*). Since the mercury excess in these plants is only partially metabolized into compounds, samples collected for analysis should never be dried. The highest uptake has been observed in coniferae (*15*). Mercury content in vegetation samples shown in Table VI ranges from 0.2 to 10 μg/gram. These values were obtained by wet ashing of dried samples and calculated on a dry basis. Because of the loss by drying as previously explained, the values are presented for illustrative purposes only. On another plane, coniferae litter (*73*) accumulates gaseous mercury in preference to organically complexed metal; therefore, litter collected near decaying sulfides has a greatly enhanced content of the metal in comparison with the underlying clay.

Reducing soils (negative redox values) may have free sulfur formed by anaerobic transformation or from hydrogen sulfide decay near volcanic active zones or be reducing just because of excess organic matter content. The mercury in these soils is firmly held by physical and chemical forces, very likely as the insoluble sulfide or some insoluble organic complexes. Besides, the solubility and volatility of the mercuric sulfide are greatly depressed when the compound is in presence of free sulfur or soluble sulfides at neutral pH (*78*). Therefore, the mercury available to plants is very low. Plants growing under these conditions take up very small amounts and, as shown in Table VI, exhibit concentrations typically in the range of 0.01 to 0.04 μg/gram.

Animals. Some information on the uptake and elimination of mercury by animals is confusing and complicated by analytical uncertainties, unrealistic experimental conditions and biased interpretations. The following is a discussion of some of the more trustworthy information available (*4, 5*). Several groups of mercury compounds can be considered: metallic gaseous, inorganic, alkoxyalkyl, aryl, alkyl, dialkyl, S and N-protein bound of all above compounds, non-lipotropic compounds containing strong C–Hg bonds with or without radicals conferring water solubility (*79*). Each compound behaves differently in living organisms (*35*). Although mercury concentrates in the food chain, animal studies indicate that the body burden is in balance with the source through intake, elimination, and time span (*2*). The elimination rate is a function of species, compound, and route of entry (*5*).

Studies performed on fish showed an increase of the levels in specimens caught in some rivers from 1940 on (*80; cf. 125*). This is in accordance with the beginning of large scale industrial operations and disposal of large amounts of waste water in river streams. On the other hand, levels of mercury in fish caught in the oceans have not changed to a measurable degree between 1878 and today (*81*). Therefore, in considering uptake of mercury by aquatic biota a distinction between the relatively uncontaminated ocean and contaminated aquatic freshwater media must be made. In the world's oceans, mercury is concentrated through the natural food web, zoo- and phytoplankton being the first link. Organisms subsisting only on plankton concentrate the mercury by factors of 3 to 6 and may have levels in flesh comparable in magnitude with those of land animals (*2, 27*). Predators accumulate higher levels because of another concentrating step. The accumulation is not linearly related with the concentration in food and cannot be extrapolated. Moreover, biochemical reactions and long retention times lead to the accumulation of lipotropic mercury compounds. Therefore, almost all of the mercury found in oceanic animals occurs as organomercurial compounds.

Eventually, all mercury entering into the ocean ends up in sea sediments. Because higher than background levels were found in sediments near zones of active seafloor spreading (with a few values up to 0.4 μg/gram), the idea of an active underground injection of mercury around these rift areas was postulated (*82*). The concept was jeopardized by the possibility of a faster turnover of biota because natural upwelling of water, conditioned by the bottom topography over these rift areas, brings nutrient salts of biogenic importance to the oceanic surface (*65*). Later analyses showed no apparent geographical trends in the distribution of mercury on deep-sea sediments, the average being higher (mean 0.41 μg/gram) than previously reported values. Furthermore, evidence that all offshore sediments including estuary, shelf, and deep-sea sediments contain about the same concentration of mercury was presented (*83*). The difference between both reports was ascribed to variables introduced when analyzing and when calculating the values. The error during analysis was introduced by heating sea sediments to only 500°C. This temperature was insufficient for total release of mercury, and a temperature of 900°C gave values close to the maximum content. Calculation of mercury content was different in both cases. In the former case, the mercury contained in the carbonate fraction of the sediment was subtracted before comparisons between sediments were made. Both studies revealed the existence of heat stable compounds of mercury in deep-sea sediments. Heat stable compounds have also been observed by the author in some rocks and sediments (*cf.* pyrophosphates and pyroarsenates in Table I).

The high levels of alkyl mercurials found in predators feeding on contaminated food were the reasons for increased research on these mercury compounds in living organisms. Inorganic mercury as well as alkyl mercurials are taken up readily by aquatic plants (*68, 84*). Phyto- and zooplankton concentrated both soluble inorganic and alkylated mercury compounds in seawater over 100 times (*1, 64, 65, 66, 85*), which is not high when compared with other elements. Inorganic mercury is transformed into the more toxic alkyl mercurials. Methylmercury compounds are produced from mercury contained in sediments by methylating bacteria (*86, 87*) favored by low pH (*32*) and nearly anaerobic conditions (*3*). The methylating process involves principally B_{12} in a non-enzymatic fermentation (*6, 37, 69, 74, 87, 88, 89*), which may occur also inside living organisms (*90*). Under the conditions studied, the methylmercury compounds so produced are in equilibrium with 40% of the mercury tied up as the biologically inert dimethylmercury. The equilibrium shifts toward methylmercury compounds at lower pH (*91*). The highest rate of production is observed when sediments have levels above 1 μg/gram (*69*).

Alkyl mercurials were 10 to 100 times as toxic as soluble inorganic mercury compounds because of lipotropy and other biological interactions (*79, 92*). Lipotropic mercury compounds damage motor and sensory nerves, especially during growth (*74, 88, 93*). Whereas inorganic protein-bound mercurials are absorbed to a low degree in the intestinal tract, methylmercury derivatives are almost totally absorbed. These compounds are cumulative in the food chain to a degree which depends on the species considered and the experimental situation. Alkyl mercurials are concentrated by aquatic organisms much more effectively than non-alkylated mercury complexes; thus, fish concentrate methylmercury chloride by an overall factor of 3000 (*6, 74, 91, 93*), and shellfish concentrate it by an overall factor of 100 to 100,000 (*57*). These large factors can be explained by the long retention times which are greater than one year. In Minamata, Japan, and elsewhere, people and domestic animals became intoxicated when a large portion of their food supply was fish containing 5–20 μg methylmercury compounds per gram (*91*). Susceptible persons showed intoxication at lower levels. Using the lowest whole-blood concentration at which toxic symptoms have occurred in humans, it was calculated (*2, 74*) that intoxication corresponds to the daily intake of 300 μg mercury per day as methylmercury, for a person with an average weight of 70 kg. Using a safety factor of 10 the allowable daily intake of alkyl mercurials would be 30 μg mercury per 70 kg weight. This number corresponds to the intake of 3 kg of fish with a level of 0.01 μg/gram, which seems to be a level difficult to find in seafood in the light of recent analyses. If fish containing 0.5 μg/gram were eaten daily, only 60 grams could be consumed safely (*74*).

Further studies established that most alkyl mercurials pass through the bile and intestines and are reabsorbed (*94*). A logarithmic elimination curve which reaches background levels in 40 to 60 days (*50*) has been found with rats given single doses. Human volunteers excreted half in about 70 days (*74, 95*). The most active sites for mercury elimination are the kidneys, liver, hair, and feathers (*95, 96, 97*). In uncontaminated persons the kidneys have 20 times the concentration of the overall body average (*98*). Higher factors are found in persons exposed to mercury. High concentrations are also found in thyroid and pituitary glands (*99*).

It is important to know in what chemical form the metal is present (*35, 100*) because mercury can be present as compounds of different toxicity (*101*). Thus, limits arbitrarily set for mercury in food could be either too low (*102*) or too high (*103*). To illustrate this point we may consider vegetation, which in general may contain higher levels in fresh tissue than generally accepted as safe (*99*). Because most analyses were made on previously air or oven-dried samples, reported values seemed to be generally low. Man's uptake can be calculated based on an average diet composed of 55% grains, 12% meat, 1% fish, and the balance, vegetables plus fruit (*25*). For 600 grams dry weight intake per day, these sources yield about 50 μg of mercury, from which about 6 μg is in form of alkyl mercurials. The daily urinary elimination for unexposed persons is about 10 to 20 μg. Assuming no retention in long range experiments, most of the remainder must have been eliminated through feces. Since most of the mercury was ingested from the vegetables plus fruit, uptake from vegetation may be low (*94*).

In biogeochemistry (*104*) high death rates occur in areas under conditions of trace element deficiencies (*88*), which may enhance adverse effects of other toxicants (*27*). Development of tolerance which leads eventually to dependence has been well documented for several trace elements (*27, 77*), but for mercury compounds the available information is inconclusive (*96*). However, the wide variation of toxicity of mercury compounds in different animals seems to be related to the simultaneous intake of naturally occurring selenium (*105*) and perhaps iodine (*99*). In a typical experiment the toxic symptoms of higher level of alkyl mercurials were reduced substantially when replacing a corn–soya diet by fish containing natural selenium (*106, 107*). Extrapolation of this effect to humans remains to be demonstrated and would give a clue about tolerance to methylmercury of some populations. Since total removal of mercury from the dietary intake would be almost impossible, deficiency of selenium in food should be controlled at the same time for establishing safety values for both elements simultaneously and for decreasing the danger of mercury intoxication.

The Mercury Cycle

The cycle depicted in Figure 1 is based on information presented in this review. Some points missed in the text are clarified below. Most data on biota production was extracted from Ref. *25*. Other information about the earth's parameters is summarized in Table VIII. The mercury in transport from source to sink is given in metric tons per day and should be accurate within a factor of 10. The mercury content of the deep sea and coastal water includes the entire body of water. The land and coastal sediments include only the top 1000 m. Because of ocean floor spreading, the accumulated deep sea sediments are highly variable. Four hundred meters should be a reasonable mean value for their thickness based on the drill core logs from the oceanographic ship *Glomar Challenger* (*108*). The amount contained in deep sea sediments has been calculated as follows:

$$\text{amount} = \text{ocean floor surfaces} \times \text{thickness of sediment} \times \text{density} \times \text{concentration}$$

Substitution yielded:

$$\text{amount} = (35 \times 10^7 \text{km}^2 \times 10^6 \text{ m}^2/\text{km}^2)\ (400 \text{ m})\ (2.5 \text{ tons/m}^3)\ (0.4 \text{ ppm} \times 10^{-6} \text{ tons/ppm})$$

$$= 14 \times 10^{10} \text{ tons}$$

For calculations soil mercury is that found to the depth of 1 m while alluvium has been considered as the 10-cm layer covered temporarily by fresh water. The contents in soil and alluvium have been calculated analogously to the example for deep sea sediments.

Atmospheric mercury is adsorbed on particulate matter and then carried down by rain (10^{14} m^3 per year) as explained in the corresponding section. Reported concentrations of mercury in rainwater are highly variable because most samples were collected near urban areas (Table VII) (*1*). As a base for calculations, a value of 0.1 μg/liter has been assumed (*58*). The amount of mercury in precipitation per day would be:

$$\text{amount per day} = \text{volume of precipitation per year} \times \text{concentration}/365 \text{ days}$$

$$= (10^{14} \text{ m}^3/\text{y})\ (0.1\ \mu\text{g/l} \times 10^3 \text{ l/m}^3 \times 10^{-12} \text{ ton}/\mu\text{g})/365 \text{ days}$$

$$= 28 \text{ tons/day}$$

The amount presented in the cycle has been rounded off to 30 tons/day. The rainwater running over land surface exchanges its mercury with that in soil and establishes an equilibrium before draining into creeks and rivers. About 20 tons/day in immediate runoff is considered adsorbed on suspended particles whereas 10 tons/day can be considered as dissolved (section on Alluvium and Internal Waterways). The immediate runoff partially evaporates, percolates, and exchanges its mercury with alluvium and atmospheric air. Any input of mercury into the internal waterways will be subject to the same line of action. The final amount of runoff, about a third of the precipitation on land (*109*), enters into coastal waters. A portion of the overall input of mercury into the environment will be also translocated to the coastal waters because of the relatively long residence time in that body compared with the residence time of mercury in alluvium. The amount of mercury contained in that drainage, 14 tons/day, has been presented in the corresponding section on Alluvium. This amount can be considered split into 10 tons/day on clay particles and 4 tons/day dissolved mercury. Most of the larger particles settle out near the coast. Rainout of mercury into coastal and open seawaters has been assumed to be much lower than over land mainly because of meteorological factors (dust load, concentration of mercury in ocean air, wind speed, and rain frequency). In effect, levels of atmospheric mercury drop quickly to constant low values off the coast. Moreover, the dust load over the ocean is low, and concomitantly the rainout of mercury collected on particulate matter results in concentrations one-tenth lower than that over land. The numbers based on these speculations are tentative, and more research is desirable to assess more accurate values. Dry fallout was included in the rainout figure for coastal and open seawaters.

Information on the land biota is also tentative. Because most land plants accumulate woody parts for an extended period of their life, the overall amount of mercury in the total land biomass is proportionally greater than that in water biota. The huge amount of litter produced in forests (10^{11} tons/year) (*25*) carries a large amount of the immobilized mercury to the forest floor. From there, it is incorporated into humus and finally into the soil. For this reason the uppermost soil (A_0 and A_1 horizons) is enriched in mercury by a factor of 2 to 4 with respect to that in underlying layers. To a smaller extent the same situation applies to agricultural land and pastures. The removal of mercury from soil by cropping represents barely 5% in 1000 years, which means that agricultural soil is really not in equilibrium with respect to natural soils.

Because of the limited information on mercury concentration in animal biomass, many of the estimates are tentative. The uptake by land animals can be considered as being nearly the same magnitude (*27, 98*) or somewhat less than that for salt water fish (uncontaminated, Table VII). However, because of the shorter half-life of the cycle (1 to 6 months) compared with the half-life in fish (1 to 2 years), the total mercury content in the annually produced land animal biomass may amount to only about 10 tons. A tenth of that value is contained in domestic animals. Man represents less than 2% of land animal biomass production.

For the animal biomass production in the oceans, which represents about 1 to 5 $\times$ 10^9 tons/year (*25*), a total mercury content of about 50

Table VII. Mercury

	Concentration of Mercury in Water, μg/liter
Rain	0.01–0.48 (*1*)
Ice prior to 1952	0.06 ± 0.02 (*40*)
Ice from 1960 to 1965	0.13 ± 0.05 (*40*)
Spring (running through sandstone with 0.2 μg/g)	0.05–0.13 [a]
Spring (mercury mine drainage, The Geysers, Calif.	0.04 [a]
River water (uncontaminated, Italy, Germany)	0.01 to 0.05 (*1*, *43*)
River water (U.S., taking off highest values)	0.04–0.5 [b]
Lake Huron	[d]
Lake Erie	[d]
Lake St. Clair	[d]
Niagara	0.06 (*118*)
Ocean, uncontaminated	0.03 to 0.27 (*1*)
Japan, Minamata Bay	1.6–3.6 (*66*)
Sweden, lakes	[d]

[a] Author's data by method from Ref. *17*.
[b] Calculated from Ref. *1*.

tons with a half-time of 2 years can be assumed. This provides roughly a 0.1 ton/day movement through the seawater–animal cycle. The total seawater–biota cycle, however, has been estimated as being much larger —on the order of 6 tons/day. Therefore, fish would incide maximally 2% on the total seawater–biota cycle. A similar value in percent can be inferred for the freshwater cycle. The total freshwater–biota cycle has been calculated to be one-fifth of the seawater–biota cycle or 1.2 tons/day.

The global impact of man can be summarized as follows. Solid refuse (including pesticides, increased fallout, and rainout) ends up on the soil surface and will increase the evaporation rate from soil and the uptake by land biota by about 20 to 25% above the present amounts for the next 3000 years or until depletion of natural resources, provided the input continues at the present rate. Liquid wastes add an overall 12% to already circulating mercury on freshwater and alluvium bodies, the impact remaining localized for *ca.* 30 years. Because of the strong adsorption of mercury on sediments, it is likely to stay either immobile or to be diluted when covered by fresh alluvium so that the addition is only a small percentage of the existing mercury in alluvium near the point of discharge. However, continuing discharges would create a deleterious and undesirable effect to freshwater systems by increasing the level and

Content in Water and Fish

Fish Type	*Concentration of Mercury in Fish, μg/g*
[d]	[d]
[d]	[d]
[d]	[d]
[d]	[d]
[d]	[d]
[d]	[d]
[d]	0.08–0.5 [c]
freshwater fish	0.05–0.15 (*119*)
	0.11–2.0 (*3*)
	0.15–3.6 (*3*)
	0.01 (*118*)
tuna	0.03–1.0 (*23*)
swordfish	1 (*103*)
protein concentrate	0.3–0.9 (*120*)
saltwater fish	9–24 (*2*)
pike	0.04–8.4 (*2*)

[c] Calculated from Ref. *3*.
[d] Not determined.

the covered surface of mercury-containing alluvium, with the attendant aftereffects on freshwater biota.

Gaseous emissions, such as those occurring during pyrometallurgical processes and burning (fossil fuels, refuse, sludge) plus cold vapor emission from products containing mercury compounds, enter the atmosphere and cause a 20–25% increase in atmospheric rainout and fallout. The impact will be distributed over wide areas of land surface for the next few thousand years until the depletion of the natural resources (ores and fuel). It is expected that the mean soil concentration reported now as 0.07 μg/gram will about double or triple whenever the existing mineral deposits and other sources are totally depleted. This may not cause any noticeable impact since soils much higher in mercury have supported agricultural operations in extensive zones of this globe. A modest 10% increase in the concentration of continental shelf and deep sea sediments will be the effect for several thousands of years thereafter.

Because of the fast input of mercury by man not all the entities in Figure 1 are balanced. Soil accumulates 33 of the 37 tons/day, and coastal sea accumulates another 2 tons/day. The remainder has been balanced between land biota and deep sea.

Table VIII. Earth Parameters Used in the Calculations[a]

Total land surface	km^2	148×10^6
Ocean surface	km^2	361×10^6
Shelf surface	km^2	30×10^6
Lake, river, underground water	km^3	0.3×10^6
Oceanic water	km^3	1335×10^6
Ice (polar and continental)	km^3	47×10^6
Tropospheric volume (first 12 km)	km^3	6000×10^6
Precipitation on land, per year	km^3	107×10^3
Precipitation on oceans, per year	km^3	411×10^3
Evaporation from land, per year	km^3	70×10^3
Evaportation from oceans, per year	km^3	448×10^3
Runoff to oceans, per year	km^3	37×10^3
Suspended solids, per year = 13 km^3	tons	33×10^9
Dissolved solids, per year	tons	5.4×10^9
Total atmospheric mass	tons	5.2×10^{15}
Coastal erosion, per thousand years	cm	100
Sedimentaion rate in ocean, per thousand years	cm	3
Sedimentation rate on shelves, per thousand years	cm	10 to 30

[a] Information based on Refs. *58, 109, 121, 122.*

An interesting point is the apparent depletion of mercury in bedrock relative to deep sea sediments. With actual rates of erosion and deposition the residence time will not balance with the age of the earth. This may indicate that in the past the rates of erosion and deposition and the concentration on land may have been much higher than it is now. On the other hand, deep sea sediments are not the ultimate sink for mercury. According to the plate tectonics theory, these sediments will spread with the ocean floor and will be transported and thrust beneath continental margins within geologic times. As a result of that transport, coastal geosynclines are filled with shelf and deep sea sediments, causing mercury to be translocated to bedrock during emergence.

Acknowledgments

Indebtedness to Samuel H. Williston, who provided monitoring data and laboratory notebook information, is hereby expressed. Valuable comments and editorial assistance of the staff of the Air and Industrial Hygiene Laboratory have been very helpful.

Glossary[a]

Absorption: an irreversible chemical retention effect on surfaces, either liquid or solid. Improperly applied to the dissolution (physical effect) of gases in liquids.

[a] Based on Refs. *123* and *124.*

Adsorption: a reversible physical retention effect or adhesion of molecules or ions on solid surfaces or liquid interphases.

Agricultural soil: soil used for cultivation.

Alluvium: detrital deposits resulting from the erosion caused by rain and water of relatively recent time, upon land not permanently submerged beneath the waters of lakes and seas.

Anomalous: a significant departure in physical of chemical value from the mean value of a larger area.

Biota: the animal and plant life of a region; flora and fauna collectively in a smaller environment.

Biotite: a mineral, member of the mica group, of dark color. A common rock-forming mineral.

Clastic material (syn. detrital material): consisting of fragments of rocks or of organic structures that have been moved individually from their place of origin.

Continental shelf: the zone around the continents, extending from the line of permanent immersion to the depth at which there is a marked increase of slope to greater depth. Estimated to be 8% of the total ocean surface.

Differentiation: the sum or processes whereby materials separate into distinctly different substances—*e.g.*, magmatic segregation.

Dissolution: a physical effect caused by the uptake of gases, liquids, or solids into liquids as opposed to absorption.

Eh values: same as redox potential. The voltage obtainable between an inert electrode placed in an electrically conductive environment and a normal hydrogen electrode, regardless of the particular substances present in the environment.

Eolian: a term applied to the erosive action of the wind and to deposits which are the result of the transporting action of the wind.

Fallout: the descent through the atmosphere of particulate matter. Formerly referred to radioactive particles only.

Franciscan sediments: Jurassic-Cretaceous sediments occurring in California and composed of graywacke.

Global environment: the sum of all the external conditions which may act upon an organism or community to influence its development or existence on the whole Earth as opposed to smaller environments.

Granite wash: the material eroded from outcrops of granitic rock.

Granitic rock: coarse-grained igneous rock, generally containing quartz and more than 55% silica.

Graywacke: a poorly sorted sedimentary rock with subrounded and angular components and generally dark in color. The ferromagnesian equivalent of arkose.

Halo: a differentiated zone surrounding a central object—*i.e.*, an alteration around an ore body.

Humates: complexes formed between metals and humic acid.

Humic acids: acidic compounds contained naturally in soils and formed by partial decay of organic matter.

Igneous rocks: formed by solidification of molten or partially molten magma.

Lenses: a body of ore or rock shaped like a double-convex optical lens, thick in the middle and thinning out at the edges.

Lipotropic: with affinity for fat or fatty tissues.

Litter: the uppermost slightly decayed layer of organic matter on the forest floor.

Magma: molten or partially molten rock material within the earth from which an igneous rock results by cooling.

Magmatic heat: heat derived from cooling of magma.

Mineralization: the destruction of organic matter to separate the inorganic portion. Geol., the action of supplying with minerals generally by means of a mineralizing fluid.

Natural soil: the undisturbed earth material which has been so modified and acted upon by physical, chemical, and biological agents that it will support rooted plants. All unconsolidated, undisturbed material above bedrock that has been in any way altered or weathered (as opposed to agricultural soil).

Oxidation potential: a redox potential above an *Eh* of 0 volt (positive values).

Oxidizing condition of the atmosphere: coexisting with ozone and nitrogen oxides or other oxidizing substances, causing oxidation of oxidizable substances.

Pelagic organisms: related to water of the sea as distinct from the sea bottom.

Pelagic sediments: related to sediment of the deep sea as distinct from that derived directly from the land.

Precipitation: term applied to hydrology. The discharge of water, in liquid or solid state, out of the atmosphere, generally upon a land or water surface, and measured as the depth of a liquid.

Rainout: the effect of removal of a substance or particle by impaction or nucleation with rain droplets.

Reducing condition of the atmosphere: coexisting with sulfur dioxide or other reducing substances, causing absence of ozone and nitrogen dioxide.

Reducing potential: a redox potential below an *Eh* of 0 volt (negative values).

Sediment: anything settling down from suspension in water or air.

Sedimentary rock: a rock formed by compaction of sediment, generally in stratified form.

Smaller environments: any delimited area of land and corresponding space of air containing all elements amenable to natural chemical and biotic interaction; a region; *see* biota; as opposed to global environment.

Spring: a natural source of water issuing from the ground.

Sulfidic base metal: a metal chemically more active than gold, silver, and platinum metals. Commonly restricted to the ore metals, such as those from the series 5, 7, and 9 from the periodic table of the elements.

Ultrabasic rock (syn. Ultramafic): igneous rocks containing less than 45% silica, devoid of quartz and feldspar, and composed essentially of ferromagnesian silicates.

Volcanic activity: designating or pertaining to the phenomena of volcanic eruption; the explosive or quiet emission of lava, pyroclastic ejecta, or volcanic gases at the earth's surface, usually from a volcano. Sometimes referred as belonging to mountain-building disturbances.

Washout: the effect of removal of a substance by dissolution in rainwater.

Literature Cited

1. Hickel, W. J., Pecora, W. T., *U.S. Geol. Surv. Prof. Paper* **713** (1970).
2. Nelson, N. *et al.*, "Hazards of Mercury," *Environ. Res.* (1971) **4,** 1, 69.
3. Oak Ridge N H Lab: Mercury in the Environment, AA R 4-79, Contract W 7405, Eng. 26, 1971.
4. Rehfus, R. A., Priddy, A. H., Barnes, M. E., "Mercury Contamination in the Natural Environment," (a cooperative bibliography), U.S. Dept. Interior, Office of Library Services, Washington, D. C., 1970.
5. Friberg, L., Vostal, J., "Mercury in the Environment," The Chemical Rubber Co., Ohio, 1972.
6. Jonasson, I. R., *Geol. Surv. Can. Pap.* **70–57** (1970).
7. Williston, S. H., private communication (1971).
8. Stock, A., Lux, H., *Angew, Z., Chem.* (1931) **44.**
9. Coyne, R. V., Collins, J. A., *Ann. Chem.* (1972) **44,** 1093.
10. Greenwood, M. R., Clarkson, T. W., *J. Amer. Ind. Hyg. Ass.* (1970) **31,** 250.
11. Byrne, A. R., Dermelj, M., Kosta, L., *J. Radioanal. Chem.* (1970) **6,** 325.
12. Toribara, T. Y., Shields, C. P., *J. Amer. Ind. Hyg. Ass.* (1968) **29,** 87.
13. *J. Amer. Ind. Hyg. Ass.* (1969) **30,** 195.
14. Tonani, F., *Geothermics* (1970) (2) 492.
15. Warren, H. V., Delavault, R. E., Barakso, J., *Econ. Geol.* (1966) **61,** 1010.
16. Linch, A. J., Stalzer, R. F., Lefferts, D. T., *J. Amer. Ind. Hyg. Ass.* (1968) **29,** 79.
17. Kothny, E. J., *J. Amer. Ind. Hyg. Ass.* (1970) **31,** 466.
18. Sandell, E. B., "Colorimetric Determination of Traces of Metals," Interscience, New York, 1959.
19. Kosta, L. *et al.*, "Uptake of Mercury by Living Organisms and Its Distribution as a Result of Contamination of the Biosphere in Characteristic Areas with Special Reference to Forage and Food," Annual Research Progress Report Inst. Josef Stefan, Ljubljana, Yugoslavia, 1971.
20. Magos, L., Cernik, A. A., *Brit. J. Indust. Med.* (1969) **26,** 144.
21. Magos, L., *Analyst* (1971) **96,** 847.
22. Gage, J. C., Warren, J. M., *Ann. Occup. Hyg.* (1970) **13,** 115.
23. de Goeij, J. J. M., *Chem. Eng. News* 6 (June 7, 1971).
24. Kothny, E. L., *R. Ass. Geol. Argentina* (1970) **25,** 63.
25. "The Biosphere," Freeman and Co., San Francisco, 1970.
26. Williston, S. H., *J. Geophys. Res.* (1968) **73,** 7051.
27. Goldwater, J. L., *Sci. Amer.* (1971) **224,** 15.
28. Joensuu, O. I., *Science* (1971) **172,** 1027.
29. Hauskrecht, I., Hajduk, J., *Biologia (Bratislava)* (1966) **21,** 676.
30. *Science News* (1971) **99,** 280.
31. Turney, W. G., *J. Water Pollut. Contr. Fed.* (1971) **43,** 1427.
32. "C.E.K.," *Environ. Sci. Technol.* (1970) **4,** 890.

33. Foote, R. S., *Science* (1972) **177**, 513.
34. Jacobs, M. B., Goldwater, L. J., *Arch. Environ. Health* (1965) **11**, 582.
35. Byrne, A. R., personal communication (Inst. Josef Stefan, Ljubljana, Yugoslavia), 1972.
36. "Threshold Limit Values of Substances in Workroom Air," adopted by the American Conference of Governmental Industrial Hygienists for 1972 (P. O. Box 1937, Cincinnati, OH-45201).
37. Wood, J. M., *Environment* (1972) **14**, 33.
38. Murgatroyd, R. J., *Phil. Trans. Royal Soc. London* (1969) **A265**, 273.
39. Tonani, F., personal communication (University of Florence, Italy), 1972.
40. Weiss, H. V., Koide, M., Goldberg, E. D., *Science* (1971) **174**, 692.
41. Brar, S. S. *et al., J. Geophys. Res.* (1970) **75**, 2939.
42. Dams, R., Robbins, J. A., Rahn, K. A. *et al., Anal. Chem.* (1970) **42**, 861.
43. Dall'Aglio, M. and Gragnani, *Atti Soc. Tosc. Sci. Nat. Mem. Ser. A* (1966) **73**, 553.
44. Atkins, D. H. F., Eggleton, A. E. J., *Proc. Symp. Nucl. Tech. Meas. Contr. Environ. Pollut. (Saltzburg)* (Dec. 26, 1970).
45. Warren, H. V., Delavault, R. E., *Oikos* (1969) **20**, 537.
46. Dvornikov, A. G., *Dokl. Akad. Nauk SSSR* (1968) **178**, 446; *CA* **68**, 80357p.
47. James, C. H., *Imperial Coll. Sci. Technol., Tech. Comm.* (1962) 41.
48. Mitsuyoshi, U., Komi, K., *Chishitsu Chosasho Geppo* (1971) **22**, 293; *CA* **75**, 142745 u.
49. Ozerova, N. A., Laputina, I. P., Aidinyan, N. K., *Vop. Odnorodnosti Neodnorodnosti Miner.* **1971**, 188; *CA* **77**, 91088p.
50. Norseth, T., Clarkson, T. W., *Arch. Environ. Health* (1970) **21**, 717.
51. Sumino, K., *Kobe J. Med. Sci.* (1968) **14**, 115, 131.
52. Strohal, P., Huljev, D., "Mercury-Pollutant Interaction with Humic Acids by Means of Radiotracers," *Nuc. Tech. Envir. Poll. Proc. Symp. 1970, IAEA, Vienna, Austria.*
53. Shcherbakov, V. P., Dvornikov, A. G., Zakrenichnaya, G. L., *Dopov. Akad. Nauk Ukr. RSR, Ser. B* (1970) **32**, 126; *CA* **73**, 17513z.
54. *Cal. Geo.* (1971) **24**, 132.
55. Dall'Aglio, M., *Atti Soc. Tosc. Sci. Nat. Mem. Ser. A* (1966) **73**, 577.
56. Smith, J. D., Nicholson, R. A., Moore, P. J., *Nature* (1971) **232**, 393.
57. "Man's Impact on the Global Environment," p. 137, Massachusetts Institute of Technology, Cambridge, 1971.
58. Jenne, E. A., "Mercury in Waters of the United States, 1970–1971," U. S. Geological Survey, Open File Report (1972).
59. *Kosmos* (1971) **67**, 292.
60. Rex, R. W., Syers, J. C., Jackson, M. L. *et al., Science* (1969) **163**, 277.
61. Syers, J. K., Jackson, M. L., Perkheiser, V. E. *et al., Soil Sci.* (1969) **107**, 421.
62. Rex, R. W., Goldberg, E. D., *Tellus* (1958) **10**, 153.
63. Rex, R. W., private communication (Inst. of Geophysics and Planetary Physics, Univ. of Calif., Riverside), 1972.
64. Fuyita, M., Hashizume, K., *Chemosphere* (1972) **1**, 203.
65. Klein, D. H., Goldberg, E. D., *Environ. Sci. Technol.* (1970) **4**, 765.
66. Hosohara, K., *Nippon Kagaku Zasshi* (1961) **82**, 1107; Hosohara, K. *et al., Nippon Kagaku Zasshi* (1961) **82**, 1479.
67. "Encyclopedia of Chemical Technology," Kirk-Othmer, Eds., Wiley, New York, 1967.
68. Dolar, S. G., Keeney, D. R., Chesters, G., *Environ. Lett.* (1971) **1**, 191.
69. Jensen, S., Jernelow, A., *Nature* (1969) **223**, 753.
70. Becker, G. F., *U. S. Geol Surv. Monograph* **13** (1888).
71. Dickson, F. W., Tunell, G., *Amer. J. Sci.* (1958) **256**, 654.

72. McCarthy, Jr., J. H., Vaughn, W. W., Learned, R. E. *et al., U. S. Geol. Surv. Circ.* **609** (1969).
73. Trost, P. B., Bisque, R. E., *Proc. 3rd Intern. Geochem. Symp., Toronto, April 1970* (Canadian Inst. of Mining and Metallurgy).
74. Dunlap, L., *Chem. Eng. News* 22 (July 5, 1971).
75. Byrne, A. R., Kosta, L., *Vestnik SKD* (1970) **17,** 5.
76. John, M. K., *Bull. Environ. Contam. Toxicol.* (1972) **8,** 77.
77. Stahl, Q. R., "Preliminary Air Pollution Survey of Mercury and its Compounds," Litton Systems Inc., 1970.
78. Rinse, J., *Rec. Trav. Chim.* (1928) **47,** 28.
79. *Ann. N. Y. Acad. Sci.* (1957) **65,** 357–652.
80. Evans, R. J., Bails, J. D., D'Itri, F. M., *Environ. Sci. Technol.* (1972) **6,** 901.
81. *Chem. Eng. News* 34 (Aug. 30, 1971).
82. Bostrom, K., Fisher, D. E., *Geoch. Cosmoch. Acta* (1969) **33,** 743.
83. Aston, S. R. *et al., Nature (Phys. Sci.)* (1972) **237,** 125.
84. *Chem. Eng. News* 34 (Sept. 27, 1971).
85. Ishikura, S., Shibuya, C., *Eisei Kagaku* (1968) **14,** 228.
86. *Chem. Engr.* 45 (Dec. 19, 1966).
87. *Nature* (1971) **229,** 3.
88. *Chem. Eng. News* 29 (July 19, 1971).
89. Wood, J. M., Kennedy, F. S., Rosen, C. G., *Nature* (1968) **220,** 173.
90. Imura, N., Pan, S. K., *Chemosphere* (1972) **1,** 197.
91. Abelson, P. H., *Science* (1970) **169,** 3.
92. Plunkett, E. R., "Handbook of Industrial Toxicology," Chemical Publishing Co., New York, 1966.
93. *Chem. Engr.* 84 (July 27, 1970).
94. Norseth, T., Clarkson, T. W., *Arch. Environ. Health* (1971) **22,** 568.
95. Birke, G. *et al., Arch. Environ. Health* (1972) **25,** 77.
96. Byrne, A. R., Dermelj, M., Kosta, L., *Nucl. Tech. Environ. Pollut. Intern. At. Energy Agency* (1971) **SM-142 a/24,** 415.
97. Berg, W. A. *et al., Oikos* (1966) **17,** 71.
98. Joselow, M. M., Goldwater, L. J., Weinberg, S. B., *Arch. Environ. Health* (1967) **15,** 64.
99. Kosta, L. *et al.*, "Fate and Significance of Mercury Residues in an Agricultural Ecosystem," Joint FAO/IAEA Research Coordination Meeting, Euratom Centre, Ispra (Italy), Nov. 1972.
100. *Chem. Eng. News* 40 (July 5, 1971).
101. Schroeder, H. A., *Science News* (1971) **100,** 63.
102. *Air Water News* (1971) **5,** 5.
103. *Chem. Eng. News* 11 (May 17, 1971).
104. "Environmental Geochemistry in Health and Disease," *Geol. Soc. Amer., Memoir* **123,** 131 (1971).
105. Parizek, J., Ostadalova, I., *Experientia (Basel)* (1967) **23,** 142.
106. Ganter, H. E. *et al., Science* (1972) **175,** 1122; *see also Chem. Eng. News* 39 (March 20, 1972).
107. *Chem. Eng. News* 16 (May 1, 1972).
108. *Geotimes* (1969–72) **14, 15, 16, 17.**
109. Debski, K., "Continental Hydrology," U. S. Dept. of the Interior and National Science Foundation, Washington, D. C., 1965, 1966.
110. Jepsen, A. F., ADVAN. CHEM. SER. (1973) **123,** 81.
111. Aidinyan, N. K., Ozerova, N. A., *Sovrem. Vulkanizm* (1966) **1,** 249; *CA* **66,** 31061n.
112. Joselow, M. M., Goldwater, L. J., Alvarez, A. *et al., Arch. Environ. Health* (1968) **17,** 39.
113. Saukov, A. A., "Geochemie," Verlag Technik, Berlin, 1953.

114. Battelle Memorial Institute, CPA Contract 22-69-153 (Aug. 1970).
115. Bowen, H. J. M., *Analyst* (1967) **92**, 124.
116. Standard Reference Material 1571, National Bureau of Standards (Oct. 1971).
117. Andren, A. W., Hariss, R. C., *Environ. Lett.* (1971) **1**, 231.
118. Kodak advertisement (1970).
119. Rotschafter, J. M., Jones, J. D., Mark, Jr., H. B., *Environ. Sci. Technol.* (1971) **5**, 336.
120. Beasley, T. M., *Environ. Sci. Technol.* (1971) **5**, 634.
121. Kuenen, P. H., Wiley, J., "Marine Geology," New York, 1950.
122. "Handbook of Chemistry and Physics," 48th ed., The Chemical Rubber Co., Cleveland, 1962.
123. "Glossary of Geology," American Geological Institute, Washington, 1960.
124. Webster's Third International Dictionary, unabridged edition, Merriam Co., 1959.
125. Harris, E. J., Karcher, R. W., Jr., *The Chemist* (1972) (5) 176.

RECEIVED January 7, 1972.

5

Measurements of Mercury Vapor in the Atmosphere

ANDERS F. JEPSEN

Environmental Measurements, Inc., Box 162, Route 2, Edgewater, Md. 21037

Measurements were made to demonstrate that a portable detector could be used to detect anomalous concentrations of elemental mercury in the air near sites where mercury had been reported in water, sediment, and fish. The Barringer airborne mercury spectrometer was installed in a car, bus, boat, or helicopter. Measurement traverses were made to locate, quantify, and map the anomalous mercury plumes. The program resulted in discovering several previously unreported sites of mercury contamination associated with natural deposits such as mercury mines and hot spring areas, and cultural sources such as chemical plants, sewage treatment plants, and sanitary land fills. Some plume dispersal patterns were over residential and populated business areas. The measurements ranged from 0 to 28,000 ng/meter^3.

A program was carried out to determine if real time measurements of elemental mercury concentrations in the air, obtained with equipment designed for mineral exploration purposes, could lead to the detection and study of mercury contamination in water by detecting anomalous elemental mercury levels in the air above.

The objectives of the program were assessing the applicability of the Barringer airborne mercury spectrometer (BAMS) to the rapid and efficient measurement of elemental mercury concentrations in air at the submicrogram/cubic meter level and relating the airborne mercury concentration measurements to mercury pollution in water.

As the program evolved it became clear that the technique could also be used to detect elemental mercury plumes in the air from ground-sited sources. Therefore, a further objective became detecting specific sources of airborne elemental mercury plumes which might contribute to the mercury load of water or sediment elsewhere.

Instrumentation

The equipment used was a Barringer airborne mercury spectrometer (BAMS, Figure 1), an atomic absorption spectrophotometer specifically designed and built to isolate the 2536.5-A emission and absorption spectrum characteristic of atomic mercury vapor. The equipment was originally developed for mineral exploration purposes and for analysis of laboratory soil samples; subsequent design improvements led to the rapid response time (1 sec) and high resolution (nanogram per cubic meter) required for airborne use (*1*).

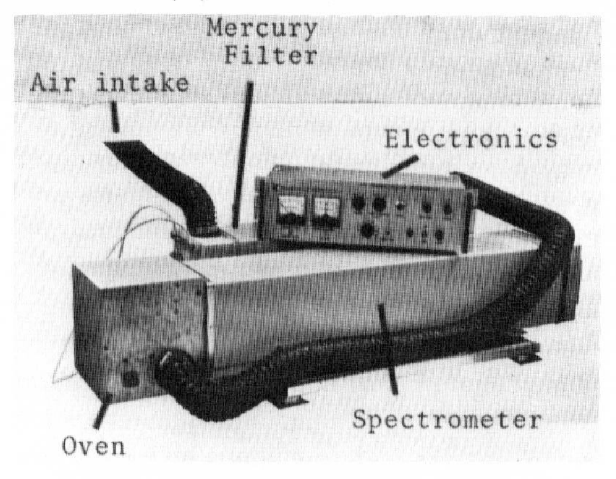

Figure 1. Barringer airborne mercury spectrometer (BAMS)

A system block diagram is shown in Figure 2. The air to be sampled is drawn through one side of a bivalve assembly into the sampling chamber at the rate of 1/3 meter^3/min; by operator choice the air passes either around or through a palladium chloride saturated filter. The filter removes elemental mercury from the air to permit determination of the instrument zero. Extensive tests have demonstrated the efficiency of this filter's specific absorption of elemental mercury to be in excess of 96%.

The source of energy is a commercial mercury neon lamp. It is operated at an elevated temperature to broaden the light spectrum; visible light is removed by an optical filter. The total light path length in the 1-meter chamber is 6 meters. Measurement of the specific absorption of the mercury in the air in the cell is detected by sequentially oscillating a saturated cell of mercury and a narrow-band interference filter alternately back and forth in front of the light. The mercury cell absorbs all of the transmitted light in the emission line, leaving the energy in the broadened edges of the spectrum. The mercury contained in the air sample in the chamber absorbs a portion of the light in the emission line. In each case, the energy transmitted is monitored by a photodetector. The difference

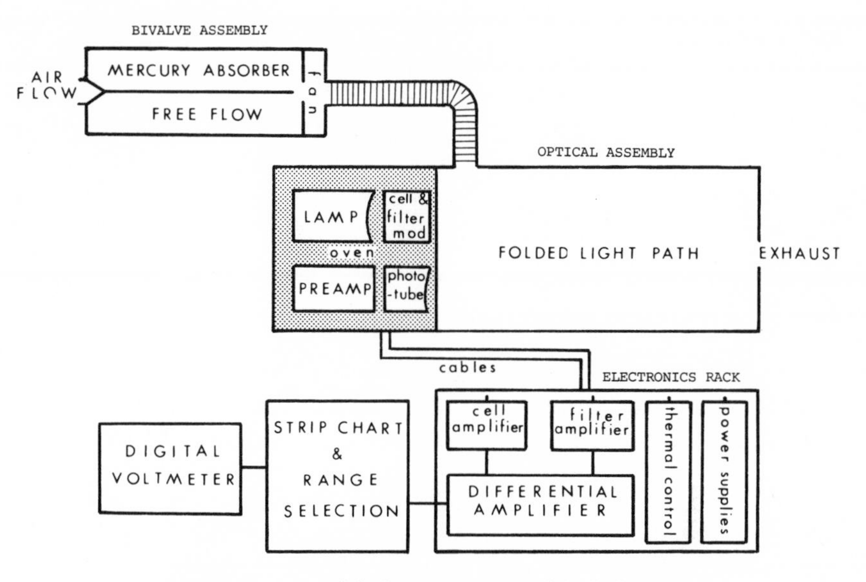

Figure 2. Simplified BAMS system block diagram

is directly proportional to the concentration of mercury present in the sample chamber. This value is amplified and recorded on a strip-chart recorder.

The temperature of the lamp and mercury cell are strictly controlled by a double-oven system at the end of the sample chamber.

Accurate calibration is achieved by injecting a known concentration of elemental mercury vapor into the sample stream and noting the response of the total detection system. The test sample is obtained by maintaining a closed bottle of mercury at a fixed temperature and drawing off a given volume, *e.g.*, 5 ml, of the saturated vapor from the air above the liquid with a syringe.

In the laboratory, the contents of the syringe are injected into the sample intake tube at a constant rate by means of a mechanized plunger. By knowing the injection rate and the system sampling rate, one can precisely calculate the mercury concentration in the air as it passes through the detection cell and hence the system response sensitivity.

In the field, calibration is done by rapidly injecting the contents of the syringe into the sample intake tube and noting the resulting deflection on the chart recorder. Agreement between the rapid injection field method and the slow mechanized laboratory method is within 95%.

Interferences caused by selective absorption by other molecules are minimized by judicious control of the thermal conditions at the lamp. The character and amplitude of the lamp's emission spectrum depend on the temperature. Interferences by sulfur dioxide and hydrocarbons which absorb at the edges of the broadened spectrum are selectively controlled by adjusting the oven temperature for minimum response to the specific interference. Thus, on highways the instrument is made to be insensitive to hydrocarbons, and at industrial areas near SO_2 sources it it made to be insensitive to SO_2. When interferences are suspected, the true mercury levels are obtained by frequent use of the palladium chloride filter to determine the instrument zero.

Field Procedures

The objectives of the program were to obtain reconnaissance field data; therefore, the equipment was installed in traversing laboratories. For work on land, the BAMS was installed in a Volkswagon microbus or a private automobile. Airborne surveying was carried out with the equipment installed in a Jet Ranger helicopter. To traverse over water suspected to be polluted with mercury, the instrument was transferred to a boat. Each of these modes required special operating power and different methods of air intake and measurement. Careful calibration and zeroing procedures were required to obtain reliable measurements of the metallic mercury concentration in the ambient air.

In each installation requirements of power, insulation to motion and temperature, and air intake and instrument operation needed specific consideration. Power was usually supplied by a 60-cycle, 110-V gasoline generator except in the aircraft where the 24-V aircraft battery was used. Venturi effects while sampling in motion were controlled by attaching an air baffle at the intake. Temperature stability was achieved by the use of insulation material, and the equipment was carefully mounted on an independent spring suspension to protect it from vibrations.

During the field program, sensitivity of the instrument varied from 6 to 15 ng/meter^3 per mV. This range, largely thermally dependent, reflected the aging characteristics of the lamp. Calibrations were carried out at half-hour intervals during each day of field measurements. Noise levels were usually below 5 ng/meter^3.

In reconnaissance, measurements were generally made from the vehicle while it was driven at standard highway speeds. This was sufficient to detect anomalies greater than 100 ng/meter^3. Once a mercury anomaly was detected, the direction of the wind was noted, and the plume was traced upwind to its source wherever possible. Traverses were then run at right angles to the wind to obtain profiles across the plume at increasing distances from the source.

Ambient levels and the precise measurement of anomalous peaks were obtained by parking the vehicle and measuring continuously for an hour or so, zeroing the instrument by means of the filter every few minutes.

Field Results

The elemental mercury vapor measurements were made in areas of Northern California known or suspected to be sites of mercury contamination. Early reports (*2*) cited only two small industries in California as having mercury-bearing effluents. Other sites were visited because of suspected emissions or were simply discovered during reconnaissance traversing. Thirteen specific sites of contamination were located (Figure 3). Of these, seven were clearly natural, and six were most likely cultural. Table I summarizes the data obtained at these sites.

Natural Anomalies. The naturally occurring elemental mercury vapor plumes were detected near the known mining area of New Almaden, in the Clear Lake area, and in the immediate vicinity of the thermal power development at The Geysers, Calif. In each of these areas, mercury mines are being operated or have been operated in the past, and the mercury vapor might be expected to emit from the soil overlying the natural deposits or from the tailings. Fish caught in these areas have been reported to contain anomalously high amounts of mercury (*3*).

NEW ALMADEN. At New Almaden, two distinct plumes containing peak values of 1500 ng/meter^3 were detected at specific points along the roadways adjacent to the operations of the New Almaden Mine. In general, the anomalies appeared to lie in the small valleys which make up the drainage of the mining area itself. To the east (downwind from the mines), levels which varied between 10 and 20 ng/meter^3 were detected near the Calero Reservoir.

CLEAR LAKE. On two separate days, mercury vapor anomalies were detected in the Clear Lake area. The maximum peak, 200 ng/meter^3, was measured at the site of the Sulfur Bank Mine (now closed). Levels of 150 ng/meter^3 were detected across the lake approximately 10 miles to the west near a golf course–residential area. This mercury may have been blown downwind from the old mine or been emitted by a local unknown source.

Approximately 10 miles to the east of Clear Lake is the still active Abbott Mine. A single traverse near this mine measured a peak of 470 ng/meter^3.

THE GEYSERS. The Geysers area was visited on a single day during which several of the steam wells were being vented. Because of the weather, the plumes from the steam well were prominent, and measurements were made both of the ambient level in the area and of the mercury content of individual steam vent plumes.

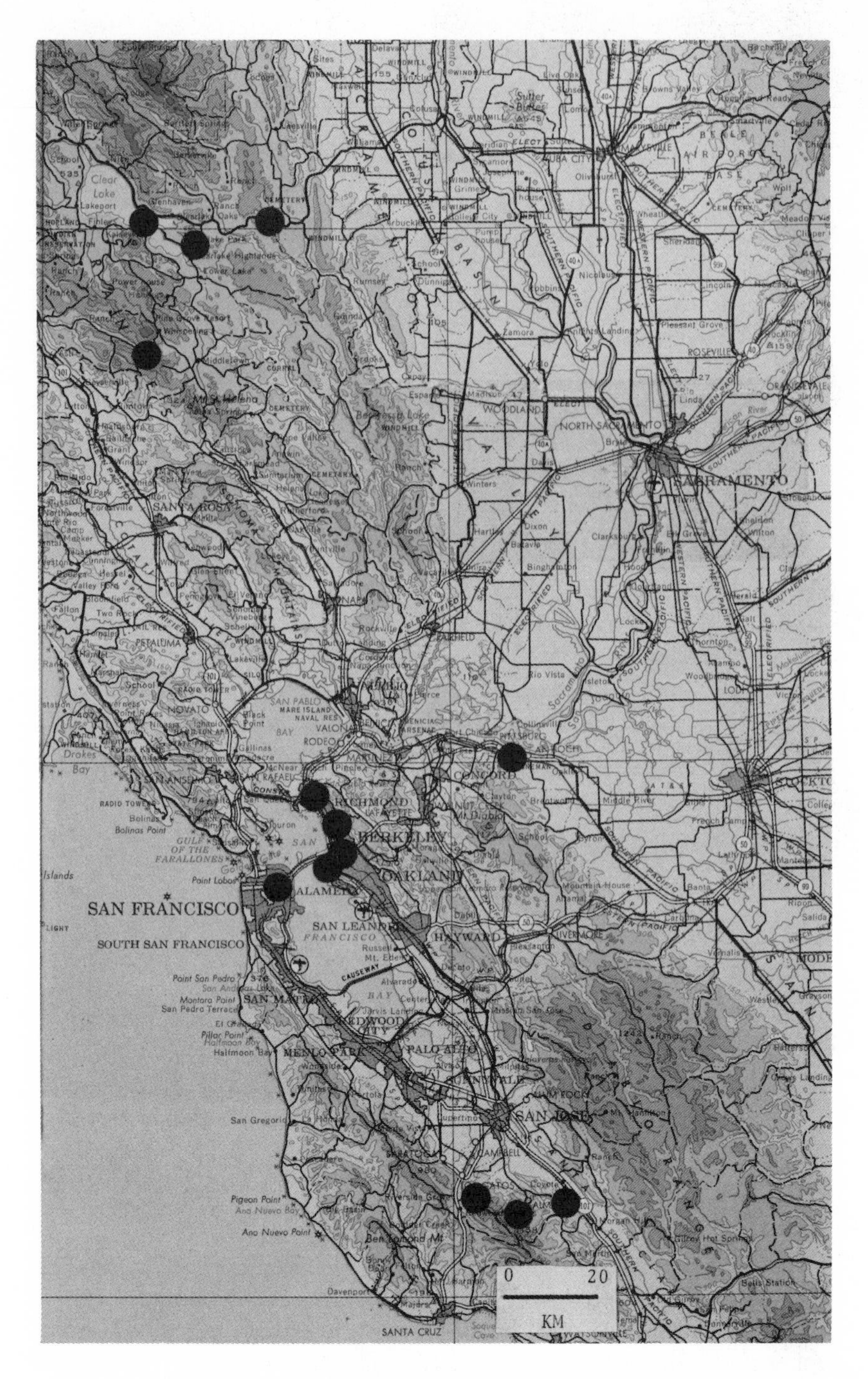

Figure 3. Locations of detected mercury plumes

The ambient levels in the valley to the west of the power development ranged from 5 to 10 ng/meter³. At the health resort at the head of the valley near the steam plants, the ambient level was 400 ng/meter³. Immediately adjacent to the geothermal steam plant and downwind of specific vents the levels varied from 200 to about 4000 ng/meter³. The maximum signal detected was in the plume emitted by a single vent on the north side of the valley; it was greater than 28,000 ng/meter³.

Cultural Anomalies. SAN FRANCISCO. Of all the sites studied, only the Quicksilver Products Co. of San Francisco had been listed as a source of mercury contamination previously. At this location, peak values between 150 and 250 ng/meter³ were detected within 100–200 feet of the site of the plant itself. These plumes were quickly dissipated and could not be traced over any significant distance.

Ambient levels in downtown San Francisco were measured continuously over two extended periods. The mercury vapor levels ranged from 0 to 100 ng/meter³. Not enough data were gathered to reveal any clear relation between the measured values and other parameters.

OAKLAND–EMERYVILLE INDUSTRIAL COMPLEX. In the heavily industrialized area in Emeryville at the east end of the San Francisco Bay Bridge, pockets of metallic mercury vapor were detected at various locations on three separate visits (Figure 4). These clouds varied in concentration from 0 to 688 ng/meter³. It was not possible to identify the source or sources of this mercury; possible local sources include a paint plant, a chemical plant, and a scrap iron recovery facility.

Downwind of a gas flare operated by the East Bay Municipal District Sewage Treatment Plant in Oakland, a maximum of 110 ng/meter³ was detected compared with an upwind ambient reading of zero. No mercury vapor was detected in the air immediately over the effluent water leaving the filtration treatment plant nor was any detected downwind of other sewage treatment plants in the area.

BERKELEY. A clearly definable mercury plume was detected downwind from a large bayfill site near the Berkeley Yacht Harbor. The maximum signal detected was 1000 ng/meter³, just upwind of highway Interstate 80. Immediately downwind of the freeway to the east, levels dropped off sharply. No more than 50–100 ng/meter³ was detected on the local streets of the residential area.

RICHMOND. A reconnaissance traverse through the city of Richmond revealed a large plume emanating from the North Richmond area (Figure 5). The maximum concentration in the plume was detected in the playground of a public school, just downwind of a pesticide and defoliant plant. Detailed traversing traced this plume for a distance of 1.5 miles southward across the Central Richmond area. A change in the wind

Table I. Summary of Atmospheric

Site	*Date*	*Time*	*Wind*
Natural sources			
Abbott Mine	Feb 12	P.M.	E
Clear Lake	Feb 12	P.M.	E
	April 2	P.M.	W
The Geysers	Feb 15	P.M.	W
New Almaden	Jan 4	P.M.	—
	March 24	P.M.	NWM
Cultural sources			
Berkeley	March 22	A.M.	W
	March 24	A.M.	W
	March 26	P.M.	W
	March 30	P.M.	W
Oakland-Emeryville	March 30	P.M.	NW
	April 5	P.M.	W
	April 6	P.M.	W
Pittsburg	Feb 11	P.M.	E
	Feb 12	A.M.	NW
	Feb 25	A.M.	N
	March 22	P.M.	WNW
	April 1	A.M.–P.M.	N
San Francisco (Quicksilver Products)	March 18	P.M.	—
	March 19	A.M.	—
San Francisco	Jan 19–26	A.M.–P.M.	—
	April 7, 8	A.M.–P.M.	—
Richmond	March 29	A.M.	NW
		P.M.	W

affected this distribution pattern so that by late afternoon it was blowing eastward directly over the residential area of North Richmond.

PITTSBURG INDUSTRIAL PLUME. During three separate trips to Pittsburg, Contra Costa County, significant mercury concentrations were detected downwind of a large chemical complex. The signal varied in amplitude and in direction depending upon the wind; a maximum of 4000 ng/meter^3 was measured. The plume appeared to be directly downwind of the settling basin which this company uses for its effluent. A local representative of the company indicated that a chlor–alkali plant had been operating directly upwind of the site of measurement and that

Mercury Measurements

Mercury Background Level, (ng/meter³)	*Mercury Peak Value, (ng/meter³)*	*Comments*
0	470	Low population desnity
0	150	Resort area, low population density
0	200	
200–800	28,100	Rural resort area
0	1,500	Rural
5–15	449	
10	800	Light industrial–residential
—	449	
10	154	
0	1,050	
0	196	Light industrial
0	688	
0	110	
	770	Industrial area
50	1,000	Industrial area
0	0	On boat
0	10	Residential
5	4,141	Industrial–residential
0		
0	278	Commercial area
0	152	
0	100	Ambient measurements, financial district
0	35	
0	1,400	Residential, near primary school
5	2,000	

it had been closed the previous year. Owing to lack of suitably located roads, it was not possible to trace the longitudinal extent of this mercury plume.

Discussion

The field data presented here are highlights from a measurement program which ranged over 30,000 km^2 and 3.5 months. Most of the measurements made during that time were of ambient levels, varying from 0 to 10 or 15 ng/meter^3, and showed no particular pattern of distribution. The anomalies detected were extremely small in lateral extent

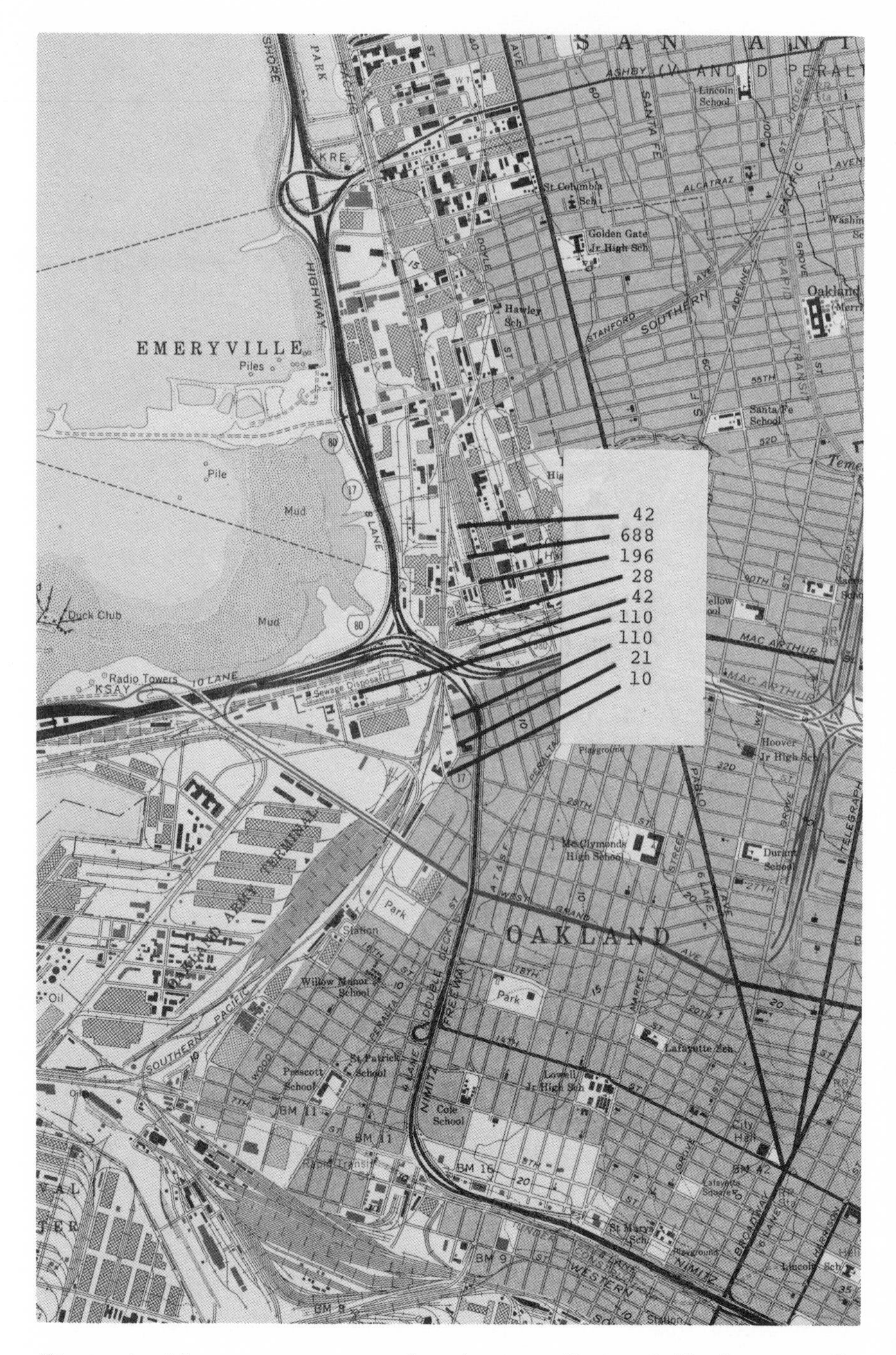

Figure 4. Mercury vapor anomalies (ng/meter³) for Oakland/Emeryville, Calif.

and extended over only limited areas. It is possible that many more anomalies still remain undetected in the area.

Only elemental mercury vapor was measured. Thus, these data alone cannot be taken as a guide to the total mercury content in the air. The rapid disappearance of the anomalous plumes may reflect not just the dispersion or deposition of the mercury but also the chemical combination of the elemental form into a combined form undetectable by the instrumentation. The rapid disappearance of the Berkeley plume as it blew across the freeway may indicate a chemical reaction catalyzed by the exhaust emissions from the freeway traffic.

A comparison of the maximum values in the natural anomalies detected in this survey is illustrated in Figure 6. By far the largest mercury concentration was detected in a steam vent at The Geysers. This signal may have been as great as it was because the measurement was made right at the source. However, the general ambient levels in the area were also much higher than elsewhere, suggesting that this area is indeed the most significant site of airborne metallic mercury. The mercury levels measured at Clear Lake may either reflect transport from The Geysers area or may arise from local sources such as the now closed Sulfur Bank mine.

The natural mercury mineralization at New Almaden is apparently responsible for most of the mercury contamination in Santa Clara County. The mercury vapor above these ore deposits has been known for a number of years (*4*). However, the route by which the mercury entered the local water reservoirs is not known. Certainly the air movement can act as a transporter of the gaseous mercury.

Figure 7 illustrates the variation in mercury peak levels around the cultural sources. These numbers reflect dispersal parameters such as wind velocity as well as the magnitude of the source.

The largest mercury vapor level attributable to an industrial source was measured near the site of the now-closed chlor–alkali plant and downwind of a settling pond used by a large chemical plant. A company representative indicated that mercury has been detected on the ground near the chlor–alkali plant site; this indeed may have been the source of the mercury vapor. However, mercury in the settling basin sediments may also be a source. If so, the 10–15 feet of water covering the sediment is clearly not an effective seal to keep the vapor in.

The measurements in Richmond indicate that the dispersal pattern of mercury vapor is similar to that of any other effluent gas. Because of the availability of roads it was possible to map the distribution pattern of this plume for over a mile.

The parameters which control the dispersion of mercury vapor plumes still require definition. No plume from an industrial source was tracked

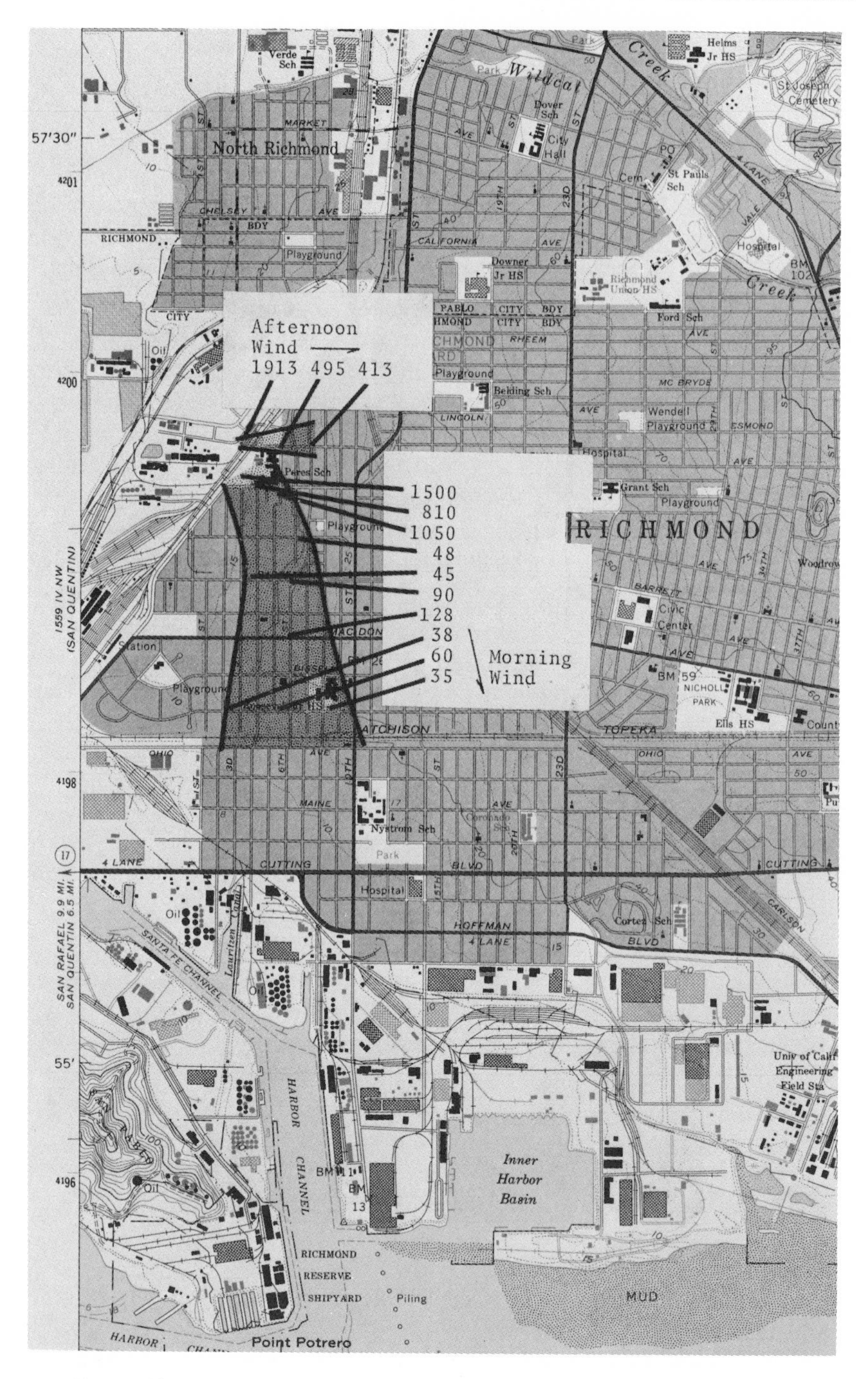

Figure 5. Mercury vapor anomalies (ng/meter³) for Richmond, Calif.

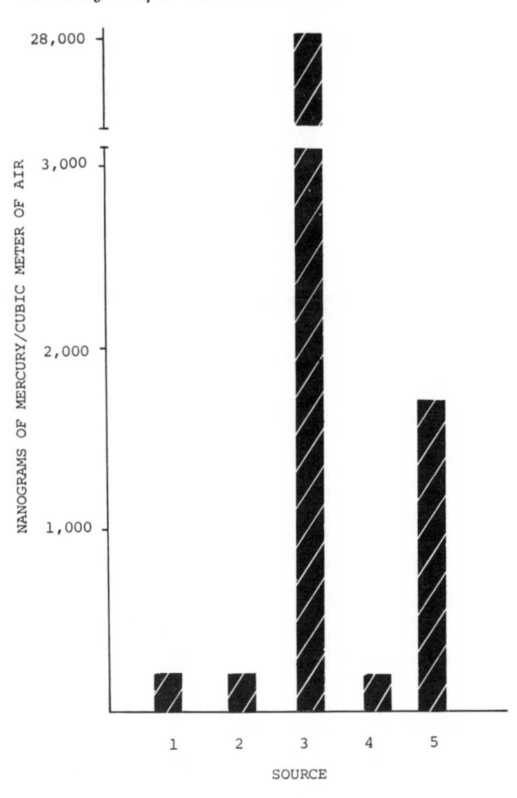

Figure 6. Peak mercury vapor levels near natural sources. 1, Clear Lake; 2, Abbott Mines; 3, Steam vent, The Geysers; 4, Coral Mine; 5, New Almaden.

for more than 1.5 miles. In Berkeley, the presence of a major well-traveled highway seemed to be instrumental in the disappearance of the mercury because of either air turbulence or exhaust emission associated with auto traffic.

Summary and Conclusion

The use of high-sensitivity, portable, rapid-response gas-analyzing equipment has led to the detection of several previously unknown elemental mercury plumes associated with both natural and industrial sources. In Northern California these anomalies were detected in a brief reconnaissance program and hence represent only a sampling of the mercury vapor plumes likely to exist in the area. An inventory of elemental mercury plumes can be obtained in any area by using a high-sensitivity, portable, rapid-response mercury detector mounted in a mobile laboratory for measurement while in motion.

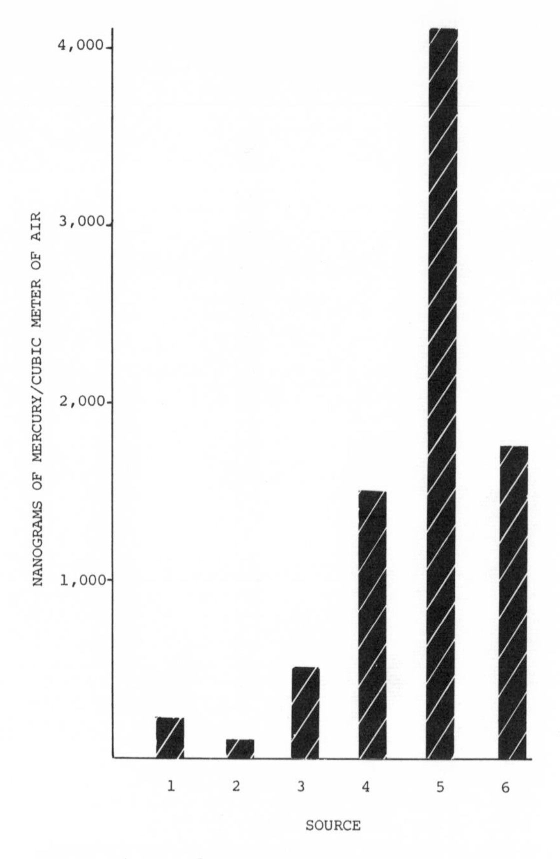

Figure 7. Peak mercury vapor levels near cultural sources. 1, San Francisco; 2, Oakland; 3, Emeryville; 4, Berkeley; 5, Pittsburg; 6, Richmond.

Both natural phenomena and human activity produce airborne elemental mercury plumes. The naturally occurring plumes are usually associated with mercury mineral deposits and appear to be enhanced by mining activity or the venting of geothermal steam. The man-made plumes are associated with such activities as industry, sewage treatment plants, and refuse fill areas. The dispersal pattern of airborne mercury plumes depends on local conditions of ventilation and perhaps also on local concentrations of other contaminating molecules.

Acknowledgment

This project was supported in part by the Water Quality Office, Office of Research and Development, the Environmental Protection Agency.

Literature Cited

1. Barringer, A. R., "Interference-free Spectrometer for High-sensitivity Mercury Analyses of Soils, Rocks and Air," *Appl. Earth Sci.* (1966) **75**, B120.
2. Wallace, R. A.,Fulkerson, W., Shultz, W. D., Lyon, W. S., "Mercury in the Environment," ORNL NSF-ER1, Oak Ridge National Laboratory, Jan. 1971.
3. Kahn, E., Chairman, "Mercury in the Environment," compiled by the Interagency Committee on Environmental Mercury, July 1970–July 1971, California State Department of Public Health, Berkeley, Calif.
4. United States Department of the Interior, "Mercury in the Environment," *U. S. Geol. Surv., Profess. Paper* **713** (1970).

RECEIVED March 20, 1972.

6

Selenium in Our Enviroment

HUBERT W. LAKIN

U. S. Geological Survey, Denver, Colo. 80225

The primary sources of selenium are volcanic emanations and metallic sulfides associated with igneous activity. Secondary sources are biological sinks in which it has accumulated. The selenium content of black shales, coal, and petroleum is 10–20 times the crustal abundance (0.05 ppm). Seleniferous black shales are the parent materials of the widespread seleniferous soils of the western plains of the United States. When burned, coal and petroleum containing selenium give rise to a redistribution of particulate Se^0 and SeO_2. The average selenium content of U. S. coal is about 3 ppm and of petroleum about 0.2 ppm. Selenium is an essential nutrient for animals and is required at a concentration of about 40 ppb in their diet; at concentrations of 4000 ppb and above, however, it becomes toxic to animals.

Selenium in concentrations of 4 ppm or more in the diet is toxic to animals. Certain soils in the western plains of the United States were found in the 1930's to produce vegetation, including wheat, that was toxic to animals because of the selenium content of the crops. Certain species of plants were discovered by Beath and others (*1*) that selectively accumulated selenium reaching a selenium content as high as 0.9% of the dry weight of the plant. These accumulator plants are exceedingly poisonous to livestock.

The discovery in 1957 of a nutritional role for selenium by Schwarz and Foltz (*2*) has introduced a new and fascinating aspect to the geochemistry of selenium. Selenium is required in the diet of animals at a minimum level of 0.04 ppm and is beneficial to 0.1 ppm; at levels above 4 ppm it becomes toxic to animals. In fact, farmers have suffered greater losses because of selenium deficiencies than selenium toxicity. Selenium deficiencies in livestock have been recognized in the United States, Canada, New Zealand, Australia, Scotland, Finland, Sweden, Denmark, France, Germany, Greece, Turkey, and Russia (*3*). The pasturage in much of New Zealand is so low in selenium (0.03 ppm) that selenium deficiency in sheep is common. It is one of the few areas in the world where addition of selenium to animal diets is legal.

Evaluation of contamination of a given environment with selenium must be tempered with the realization that a deficiency of selenium in the diet of animals, including man, is the more probable norm than toxicity of selenium in the diet.

Selenium is erratically dispersed in geologic materials. Goldschmidt and Strock (*4*) estimated that the selenium–sulfur ratio in sulfides is about 6000 and the abundance of sulfur in igneous rocks is 520 ppm. Dividing 520 by 6000 gives a figure of 0.09 ppm for the *estimated* abundance of selenium in igneous rocks. This figure was used as the crustal abundance of selenium until 1961 when Turekian and Wedepohl (*5*) revised the sulfur content of igneous rocks downward to 300 ppm and divided that figure by Goldschmidt's selenium–sulfur ratio of 6000 to obtain an estimated abundance of 0.05 ppm selenium in the earth's crust.

Table I. Selenium in Igneous Rocks

Rock Type	*Number of Samples*	*Range, ppm*	*Average, ppm*
Russian Igneous Rocks (*6*)			
Acid	5	—	0.14
Basic and ultrabasic	10	—	0.13
Alkalic	3	—	0.10
U. S. Geological Survey Standard Rock Samples (*7*)			
Diabase	1	—	0.110
Basalt	1	—	0.103
Nepheline syenite	1	—	0.010
Granite	1	—	0.005
Grandodiorite	1	—	0.059
Andesite	1	—	0.008
Peridotite	1	—	0.022
Dunite	1	—	0.004
New Zealand Volcanic Ashes and Rocks (*8*)			
Rhyolitic pumaceous ashes	11	0.10–0.24	0.17
Andesitic ashes	5	0.60–2.64	1.48
Basaltic ashes	4	0.24–1.50	0.74
Andesitic rock	2	0.47–0.92	0.69
Basalts	2	0.17–0.38	0.22
Rhyolite	1	—	0.09

Relatively few igneous rocks have been analyzed for selenium (Table I). Sindeeva (*6*) reported an average of 0.14 ppm selenium in Russian igneous rocks on the basis of six combined samples and 15 single specimens. Brunfelt and Steinnes (*7*) found from 0.004 to 0.110 ppm in seven

U. S. Geological Survey standard rock samples. Volcanic ash flows and lava of New Zealand are rich in selenium. One may conclude that the selenium content of igneous rocks ranges from 0.004 to 1.5 ppm, that some andesitic ash flows contain unusually high amounts of selenium, and that lavas are usually higher in selenium than intrusives.

Selenium and Sulfur

The high selenium content reported by Wells (*8*) in volcanic ash flows draws one's attention to volcanism. Selenium has been found in volcanic emanations and in volcanic sulfur deposits. The gases emitted by volcanoes are the original "smoke stack polluters," and through the geologic ages they have dispersed selenium in relatively high quantities. Rankama and Sahama (*9*) estimated that this source amounted to 0.1 gram of selenium per square centimeter of the earth's surface. The selenium content of volcanic sulfur has been reported to be as high as 5.18% (*10*). Selenium has not been found, however, in native sulfur derived from sedimentary processes. This difference in the association of selenium with sulfur is explained by the physical and chemical differences of these elements. The melting points and boiling points of like forms of selenium and sulfur differ markedly—selenium is consistently less volatile than sulfur (Table II). At ambient temperatures selenium dioxide is a solid and sulfur dioxide is a gas. Thus, one would expect selenium dioxide to be carried down by rain much nearer its source of emission than sulfur dioxide. The high selenium content of Hawaiian soils is probably a reflection of the ease of removal of selenium from volcanic gases. Also, Suzuoki (*12*) found that the selenium content of condensed waters from fumaroles increases with increasing temperature of the fumarole gases whereas the sulfur content remains relatively constant.

The differences between the oxidation potentials of like forms of selenium and sulfur probably account for the separation of selenium and sulfur in the weathering cycle (Table III). Hydrogen selenide is readily oxidized to elemental selenium. Selenium dioxide is easily reduced to

Table II. Some Physical Properties of Selenium and Sulfur Compounds (*11*)

Chemical Form	*Melting Point, °C*	*Boiling Point, °C*
S^0	112–119	444.6
Se^0	170–217	684–688
H_2S	−85	−60
H_2Se	−66	−41
SO_2	−75.5	−10
SeO_2	340	Sublime 315

Table III. Oxidation Potentials of Some Selenium and Sulfur Reactions (*13*)

Reaction	Potential
$H_2S = S + 2H^+ + 2e^-$	$E_0 = -0.141$ (volt)
$H_2Se = Se + 2H^+ + 2e^-$	$E_0 = +0.40$
$S + 3H_2O = H_2SO_3 + 4H^+ + 4e^-$	$E_0 = -0.45$
$Se + 3H_2O = H_2SeO_3 + 4H^+ + 4e^-$	$E_0 = -0.74$
$H_2SO_3 + H_2O = SO_4^{2-} + 4H^+ + 2e^-$	$E_0 = -0.17$
$H_2SeO_3 + H_2O = SeO_4^{2-} + 4H^+ + 2e^-$	$E_0 = -1.15$

elemental selenium. In acid solutions sulfur dioxide reduces H_2SeO_3 to Se^0.

The high oxidation potential required to form the selenate ion results in the separation of sulfur and selenium in an oxidizing environment. Sulfur is oxidized to sulfate and is relatively mobile. Selenium is oxidized to selenite and is bound in a very insoluble basic ferric selenite and is immobile.

Selenium in Soils

The selenium content of soils ranges from 0.1 ppm (*8*) in a selenium-deficient area of New Zealand to 1200 ppm in an organic-rich soil in a toxic area of Ireland (*14*). Soils of Hawaii that contain 6–15 ppm selenium do not produce seleniferous vegetation (*15*). In contrast, soils of South Dakota and Kansas that contain less than 1 ppm selenium do produce seleniferous (toxic) vegetation (*16*).

The availability of selenium in soils to plants growing in the soils is governed by the chemical form of the selenium in the soil. Elemental selenium is a moderately stable form in soils (*17*) and is not available in this form to plants. Gissel-Nielsen and Bisbjerg (*18*) added selenium to soils in the forms of Se^0, K_2SeO_3, K_2SeO_4, and $BaSeO_4$. In a two-year study they found that for mustard plants the total uptake as a percentage of the added selenium was 0.01% Se^0, 4% K_2SeO_3, and 30% of K_2SeO_4 and $BaSeO_4$; for lucerne, barley, and sugar beet the uptake was one-third or less of that obtained with mustard. The availability of selenium to plants is a function of the pH and *Eh* of the soil as well as the total selenium content of the soils. In acid soils (pH 4.5–6.5) selenium is usually bound as a basic ferric selenite of extremely low solubility and is essentially unavailable to plants. In alkaline soils (pH 7.5–8.5) selenium may be oxidized to selenate ions and become water soluble—this form is readily available to plants.

Secondary Sources of Selenium

The organic chemistry of selenium is similar to that of sulfur. Selenium may replace sulfur in simple amino acids. Both elements are

Table IV. Selenium Content

Description	*Age*
Meade Peak Phosphatic Shale Member, Phosphoria Formation	Permian
Carbonaceous marine shales, western U.S.	
Noncarbonaceous nonmarine shales	Cretaceous
Carbonaceous nonmarine shales	
Nearshore marine shales	
Offshore marine shales	
Manning Canyon Shale:	Mississippian and Pennsylvanian
Carbonaceous, pyritic dark-colored shale, 20% of 1600-foot section	
Shaly limestone and limestone, 80% of section	

necessary nutrients for animals and perhaps for some plants, and they are commonly found concentrated in organic materials.

The selenium distributed by erosion of igneous rocks or by volcanic emanations is locally concentrated in carbonaceous deposits. Black shales, coal, and, to a lesser extent, petroleum often contain much selenium due in part, at least, to selenium introduced by biomaterials.

The selenium content of black shales (Table IV) correlates with organic carbon in these shales. Vine (*19*), in a statistical study, found better than 99.5% probability of organic carbon–selenium correlation in the Meade Peak Phosphatic Shale Member of the Phosphoria Formation. Tourtelot, Schultz, and Gill (*23*) found less than 0.5 ppm selenium in Pierre Shale containing $<0.5\%$ carbon and an average of 10 ppm selenium in those containing $>1.0\%$ carbon. Seleniferous black shales in many places are the parent material of soils producing seleniferous vegetation that is toxic to animals.

The selenium contents of two Japanese coals were 1.3 and 1.05 ppm (*24*). The selenium content of 86 samples of coal (*25*) from 55 counties in 20 states of the United States (Table V) averaged 3.3 ppm and ranged from 0.46 to 10.65 ppm. Only three values were less than 1 ppm. With a crustal abundance of only 0.05 ppm these coals contain 10–200 times as much selenium as is found in igneous rocks.

Hashimoto, Hwang, and Yanagisawa (*24*) reported an average of 0.82 ppm selenium in five samples of raw petroleum available in Tokyo. In contrast, Pillay, Thomas, and Kaminski (*25*) reported an average of 0.17 ppm selenium (0.06–0.35 ppm) in 42 U. S. crude oil samples.

Shah, Filby, and Haller (*26*) found 0.026–1.4 ppm selenium in ten samples of crude oil of which five samples were from California, three from Libya, and one each from Louisiana and Wyoming. The lowest

of Certain Black Shales

Number of Samples	Selenium, ppm Minimum	Maximum	Average	Reference
43	—	—	70	*19*
6	2	75	31	*20*
8	—	—	<0.5	*21*
14	—	—	0.7	*21*
32	—	—	0.9	*21*
53	—	—	10.0	*21*
—	—	—	18	*22*
—	—	—	<1	*22*

selenium content was found in the Louisiana oil, and the highest selenium content was found in a California oil.

The burning of these seleniferous fuels gives rise to a potential contamination of the atmosphere with selenium. Pillay, Thomas, and Kaminski (*25*) estimated that the annual release of selenium from the combustion of coal and oil in the United States is about 8,000,000 pounds. This figure is nearly six times the 1964 production of selenium in the whole of North America and four times the world production for the same year (*27*). One might expect from these figures that the industrial eastern part of the United States would be seleniferous. The fact is, however, that 65% of the forage crops in the industrial eastern part of the United States (area IIC in Figure 1) contain insufficient selenium for the growth of healthy animals (*28*). The reasons for this disparity are to be found again in the availability of selenium to plants. The selenium fallout is in the form of Se^0 or SeO_2; Se^0 is relatively insoluble and only slowly oxidized; SeO_2 forms soluble selenite salts which react with ferric oxides or hydroxides to produce, in slightly acid soils, basic ferric selenites that are very insoluble. In the acid soils of the region one would not expect the oxidation of selenite to the soluble selenate.

Selenium has many industrial uses that are limited in part by its cost of about $6.00 per pound. The recovery of selenium from flues of large fuel-burning installations might help pay the cost of cleaning the effluent of the plants and most certainly would be a conservation measure.

Selenium in Natural Waters

The safe upper limit of selenium content in drinking water is considered to be 10 μgrams/liter (*29*). The average selenium content of nine

Table V. Selenium Content of U.S. Coals (*25*)

State	*Number of Counties Sampled*	*Number of Samples*	*Selenium, ppm*		
			Low	*High*	*Average*
Alabama	3	4	2.20	8.15	5.14
Colorado	2	3	1.25	2.05	1.65
Illinois	2	2	1.05	1.97	1.51
Indiana	3	4	1.41	8.36	3.96
Iowa	1	1	1.54	1.54	1.54
Kansas	1	1	2.27	2.27	2.27
Kentucky	4	5	1.71	4.82	3.13
Maryland	1	1	1.70	1.70	1.70
Missouri	1	2	3.41	4.98	4.19
Montana	3	3	2.20	4.11	3.04
New Mexico	2	2	4.43	4.82	4.62
North Dakota	1	1	0.98	0.98	0.98
Ohio	4	4	2.64	7.30	4.62
Pennsylvania	7	11	1.35	10.65	3.74
Tennessee	1	1	4.89	4.89	4.89
Utah	2	4	1.30	2.37	1.83
Virginia	3	4	2.24	6.13	4.37
Washington	1	2	0.46	0.66	0.56
West Virginia	12	30	0.92	6.80	3.36
Wyoming	1	1	3.43	3.43	3.43
Total 20	55	86	—	—	3.36

rivers (Table VI) is given as 0.2 μgram/liter (*30*). Industrial activity has only minimal effect on the selenium content of waters. For example, the Amazon River, draining a nonindustrialized area, contains 0.21 μgram/liter of Se—as much as the average. The 0.325 μgram/liter of Se found in the Susquehanna River in a coal-mining industrial area does not differ from the selenium content found in the Mad River in a nonindustrial area. The adjacent Klamath and Mad Rivers drain similar geologic environments and contain 0.122 and 0.348 μgram/liter of Se—almost the range found for the nine rivers studied.

Lindberg (*31*) found an average of 0.11 μgram/liter of Se in eight ground and surface waters of Sweden. The minimum selenium content, 0.06 μgram/liter, was found in Stockholm tap water and the maximum selenium content, 0.15 μgram/liter, was found in the waters from two wells of rural Sweden.

Scott and Voegeli (*32*) found that the selenium content ranged from less than 1 to 400 μgrams/liter of Se in 43 samples of surface waters in Colorado. Of these 43 samples, 14 contained 10–400 μgrams/liter of Se and 11 samples exceeded the 10 μgrams/liter allowable in drinking water.

The selenium content of surface waters is a function of pH of the waters as well as of its presence in the drainage system (Tables VII and VIII). Selenium is quantitatively precipitated as a basic ferric selenite

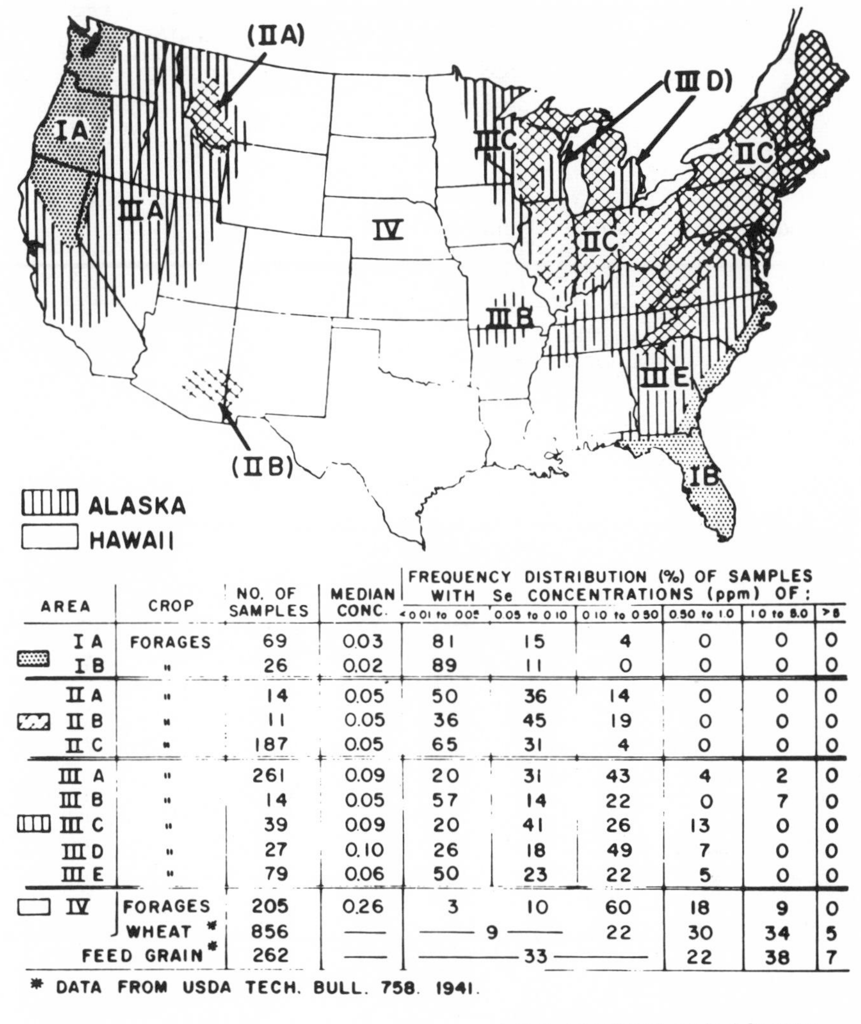

AREA	CROP	NO. OF SAMPLES	MEDIAN CONC.	FREQUENCY DISTRIBUTION (%) OF SAMPLES WITH Se CONCENTRATIONS (ppm) OF: <0.01 to 0.05	0.05 to 0.10	0.10 to 0.50	0.50 to 1.0	1.0 to 8.0	>8
I A	FORAGES	69	0.03	81	15	4	0	0	0
I B	"	26	0.02	89	11	0	0	0	0
II A	"	14	0.05	50	36	14	0	0	0
II B	"	11	0.05	36	45	19	0	0	0
II C	"	187	0.05	65	31	4	0	0	0
III A	"	261	0.09	20	31	43	4	2	0
III B	"	14	0.05	57	14	22	0	7	0
III C	"	39	0.09	20	41	26	13	0	0
III D	"	27	0.10	26	18	49	7	0	0
III E	"	79	0.06	50	23	22	5	0	0
IV	FORAGES	205	0.26	3	10	60	18	9	0
	WHEAT *	856	—	9		22	30	34	5
	FEED GRAIN *	262	—	33			22	38	7

* DATA FROM USDA TECH. BULL. 758. 1941.

Journal of Agricultural and Food Chemistry

Figure 1. Generalized regional pattern of Se concentration in crops (28)

at pH 6.3–6.7. At a pH of about 8, selenite may be oxidized to the soluble selenate ion.

The selenium content of seawater (Tables IX and X) is relatively constant in all the waters examined by Schutz and Turekian (*33*). The coefficient of variation calculated by these authors for their data is 13% based on replicate determinations. The variation about the mean of 0.09 μgram/liter for seawaters exclusive of Long Island is no greater than that given by the analytical error. The slight rise to an average of 0.11 μgram/liter in Long Island Sound may be a reflection of some contamination

Table VI. Selenium Content of Various Streams (30)

River	*Sampled at*	*Industrial Area*	*Se, μgrams/liter*
Mississippi	Minneapolis, Minn.	Yes?	0.114
Susquehanna	Marietta, Pa.	Yes	0.325
Mad	Blue Lake, Calif.	No	0.348
Klamath	Klamath Glenn, Calif.	No	0.122
Russian	California Hwy 116, Calif.	No	0.142
Eel	U. S. Hwy 101, Calif.	No	0.237
Brazos	U. S. Hwy 59, Texas	No	0.177
Rhone	Avignon, France	Yes?	0.153
Amazon	Santarem, Brazil	No	0.21
		Average of all rivers	0.2

Table VII. Selenium Content of Colorado Surface Waters, pH 6.1–6.9 (32)

Stream	*County*	*pH*	*Se, μgrams/ liter*
Animas River	San Juan	6.1	<1
Mineral Creek		6.1	<1
		6.4	<1
Animas River		6.5	<1
	LaPlata	6.9	1
Los Pinos River		6.7	<1
Animas River		6.8	<1
Vallecito Creek		6.8	<1
San Juan River	Archuleta	6.9	<1
		6.7	<1
East Fork San Juan River		6.7	1

Table VIII. Selenium Content of Colorado Surface Waters, pH 7.8–8.2 (32)

Sream	*County*	*pH*	*Se, μgrams/liter*
East Fork San Juan River	Archuleta	7.8	10
Rio Blanco		8.0	50
Navajo River		7.9	270
San Juan River		8.1	20
Hermosa Creek	LaPlata	7.9	60
Florida River		8.0	1
		8.2	400
Spring Creek		7.9	30
Animas River		8.0	40

Table IX. Selenium Content of Seawater (33)

Source	*Number of Samples*	*Se, µgrams/liter*	
		Range	*Average*
Caribbean	2	0.10–0.12	0.11
Western North Atlantic	5	0.084–0.13	0.096
Eastern North Atlantic	6	0.076–0.11	0.088
Western South Atlantic	2	0.070–0.080	0.075
Eastern Pacific	6	0.061–0.12	0.104
Antaractic	1	0.052	0.052
Long Island Sound	5	0.10–0.13	0.11

Table X. Variation of Selenium Content of Seawater with Depth of Sample (33)

Source	*Depth, m*	*Selenium, µgrams/liter*
Western North Atlantic, lat 39° 15′ N.,	5	0.11
long 63° 09′ W.	500	0.10
Eastern North Atlantic, lat 21° 21′ N.,	8	0.076
long 24° 03′ W.	600	0.088
Eastern North Atlantic, lat 9° 20′ N.,	500	0.083
long 18° 36′ W.	800	0.088
Western South Atlantic, lat 21° 49′ S.,	10	0.080
long 35° 43′ W.	100	0.070
Eastern Pacific, lat 3° 32′ N.,	14	0.080
long 81° 11′ W.	200	0.065
	400	0.12

Table XI. Selenium Content of Snowfall in the Industrial Area of Cambridge, Mass., Compared with Selenium Content of Glacial Ice in Greenland

Date of Deposition of Samples	*Number of Samples*	*Selenium Content*
Snowfall in Cambridge, Mass. (*36*)		
December 1964	2	120 µgrams/liter
December 1964 through February 1965	9	270 µgrams/liter
March 1965	5	90 µgrams/liter
Glacial Ice in Greeland (*37*)		
December 1964		9.7 µgrams/kg
Winter, 1965		14.2 µgrams/kg
Spring, 1965		8.0 µgrams/kg

from the New York City area. The contamination does not appear to us to be significant, especially as Chau and Riley (*34*) reported 0.50 μgram/liter of Se in the English Channel and 0.34 μgram/liter of Se in the Irish Sea.

Selenium in the Atmosphere

As stated previously, selenium is introduced into the atmosphere by volcanic activity and by the burning of fossil fuels, especially of coal. Lakin and Byers (*35*) found 0.05–10 ppm selenium in atmospheric dust collected on air-conditioning filters in 10 U. S. cities. More quantitative measures have been made recently using neutron activation methods to determine selenium in snow and glacial ice (Table XI). These data show that snow in the Boston area contains 10–20 times more selenium than does ice in Greenland glaciers. This variation in selenium content of snow with distance from industrial activity is in harmony with particulate selenium issuing from smokestacks.

Hashimoto and Winchester (*36*) reported an average (seven samples) of 0.09 μgram of selenium per 100 m^3 of air taken during the summer of 1965 in Cambridge, Mass. One may assume that selenium is present in the atmosphere, especially in industrial areas, and that its concentration varies with the quantity and kind of fossil fuels being burned in the area of interest.

Highly seleniferous plants have a disagreeable garlic-like odor arising from volatile selenium compounds that are released by the plants. Volatile selenium is also released by nonaccumulator plants such as alfalfa, and the amounts released are qualitatively related to the amounts of selenium within the plant (*38*). Dimethyl diselenide has been demonstrated as a volatile product of *Astragalus racemosus* (*39*).

In contrast, dimethyl selenide is given off by certain microorganisms and is exhaled by animals fed on seleniferous diets (*40*). Selenium is eliminated by humans in part through the breath (*41*) and is supposed to be in the form of dimethyl selenide. Soil bacteria expel a gaseous form of selenium (*42*). Thus, in biological processes small amounts of selenium enter the atmosphere in a truly gaseous form.

Selenium in Biomaterials

Selenium is to a certain degree concentrated by living organisms. Its concentration in organisms varies with the organism as well as with the amount available to the organism, and also varies between different parts within a given organism.

Certain species of *Astragalus* absorb selenium selectively and because of their high selenium content are always toxic to animals. Forage crops grown for animal feed in soils containing soluble selenium compounds may absorb sufficient selenium to be toxic to animals—in soils containing only insoluble selenium compounds the same forage crops may be so low in selenium that animals fed this forage will suffer from selenium deficiency.

The widespread presence of selenium in food and water and its selective absorption are illustrated by the selenium content of human blood. Allaway, Kubota, Losee, and Roth (*43*) found an average of 0.2 ppm selenium (20.6 μgrams Se/100 ml) in whole blood in 210 samples from 19 cities in 16 States in the United States. The selenium contents of these blood samples were within the narrow range of 0.1–0.34 ppm. These values for human blood are 1000-fold that found in river waters (*30*) and 2000-fold the average selenium in seawater (*33*).

Table XII. Selenium in Fish Meals (*44*)

Species of Fish	*Number of Samples*	*Selenium Content, ppm*		
		Minimum	*Maximum*	*Average*
East Canadian herring	12	1.3	2.6	1.95
Chilean anchovetta	12	0.84	2.6	1.35
Tuna	9	3.4	6.2	4.6
Smelt	6	0.49	1.23	0.95
Menhaden	12	0.75	4.2	2.09

Table XIII. Variation of Selenium Content of Menhaden Meal from Fish Caught in Different Water Areas (*44*)

Number of Samples	*Area Where Fish Were Caught*	*Selenium Content, ppm*		
		Minimum	*Maximum*	*Average*
12	Mississippi delta	1.10	3.70	1.93
12	Atlantic and Gulf	1.22	3.98	2.22
12	Atlantic	0.75	4.20	2.09

The selenium content of fish meal, shown in Table XII, illustrates the bioconcentration of selenium and the variation in bioconcentration between species. One gram of the average tuna fish meal, according to these data, contains as much selenium as does 50 liters of seawater. The selenium content shows more variation between species of fish from the same water area than between multiple samples of a single species from different waters (Tables XII and XIII).

The selenium content of six samples of fish meal from various sources analyzed by Lindberg (*31*) ranged from 1.47 to 2.45 ppm and averaged

1.94 ppm. Lunde (*45*) found an average of 2.1 ppm selenium in 11 samples of dehydrated parts of cod, herring, mackerel, and haddock. These data, combined with the data of Kifer, Payne, and Ambrose (*44*) in Table XIII, give an average of about 2 ppm selenium in fish meal.

Various algae samples from an area of industrial contamination (Trondheimsfjord) and from an area free of industrial contamination (Reine in Lofoten) contained 0.05–0.24 ppm selenium. There was more variation between species of algae than between sources of material. The low selenium in these plants as compared with the selenium content in fish is attributed by Lunde (*46*) to the different organic forms of sulfur in the plants and in fish—selenium being enriched relative to sulfur in proteins.

Summary

Selenium is unique in being both a required element in the diet for good health at a level of 0.04–0.1 ppm and a poison at levels as low as 4 ppm in the diet. In the diets of farm animals, selenium deficiency is more frequently encountered than selenium toxicity.

The crustal abundance of selenium is estimated to be 0.05 ppm. Analyzed igneous rocks contain 0.004–1.5 ppm selenium.

Selenium is associated with sulfur in volcanic activity. Unlike sulfur, selenium, upon entering the atmosphere, is most apt to be in a particulate form either as Se^0 or as SeO_2 and is subject to removal by rain relatively close to its point of origin. Thus, some sediments are relatively rich in selenium because of nearby volcanism during sedimentation.

Although the differences in physical and chemical properties of selenium and sulfur tend to separate these elements in weathering and erosion, their similar functions in biological reactions tend to unite them in organic-rich materials. Selenium is found often manyfold its crustal abundance in black organic-rich shales, in coal, and, to a lesser extent, in petroleum.

The burning of fossil fuels in the United States has been calculated to yield 8,000,000 pounds of selenium annually as a pollutant to the atmosphere. However, in the industrial northeastern United States where a large percentage of these fuels are burned, 65% of the forage crops analyzed by Kubota and his associates (*28*) contained insufficient selenium for the growth of healthy animals.

The selenium content of streams is a function of the presence of selenium in the drainage system and of the pH of the stream water. Waters in major rivers may average about 0.2 μgram/liter of selenium. Alkaline waters of southwestern Colorado that drain volcanic rock, Precambrian metamorphic rocks, and sedimentary rocks contained from 1 to 400 μgrams/liter of selenium.

The selenium content of seawater averages 0.09 μgram/liter and does not vary significantly from this average.

Selenium in the atmosphere as measured by the selenium content of snow is a function of the industrial activity in the vicinity of measurement.

Another minute source of selenium in the atmosphere is the exhalation of volatile selenium compounds by plants, animals, and microorganisms, of which the latter may be the most important. Dimethyl diselenide is a volatile given off by certain plants; dimethyl selenide is a volatile given off by microorganisms and animals.

Finally, the accumulation of selenium in biomaterials may be illustrated by the selenium content of human blood (0.2 ppm) which is 1000-fold the selenium content of surface waters, and by the selenium content of marine fish meal (2 ppm) which is 50,000-fold the selenium content of seawater.

Literature Cited

1. Beath, O. A., Draize, J. H., Eppson, H. F., Gilbert, C. S., McCreary, O. C., "Certain Poisonous Plants of Wyoming Activated by Selenium and Their Association with Respect to Soil Types," *J. Amer. Pharm. Assoc.* (1934) **23**, 9497.
2. Schwarz, K., Foltz, C. M., "Selenium as an Integral Part of Factor 3 Against Dietary Necrotic Liver Degeneration," *J. Amer. Chem. Soc.* (1957) **79**, 3292.
3. Allaway, W. H., "Control of the Environmental Levels of Selenium," in "Trace Substance Environmental Health—2, Proceedings of the University of Missouri 2nd Annual Conference, 1968," pp. 181–206, University of Missouri, Colombia, Mo., 1969.
4. Goldschmidt, V. M., Strock, L. W., "Zur Geochemie des Selens II: Nachr. Ges. Wiss. Göttingen Math.-Phys. Kl. IV," Vol. 1, pp. 123–143, 1935.
5. Turekian, K. K., Wedepohl, K. H., "Distribution of the Elements in Some Major Units of the Earth's Crust," *Geol. Soc. Amer. Bull.* (1961) **72**, 175.
6. Sindeeva, N. D., "Mineralogy and Types of Deposits of Selenium and Tellurium," Interscience, New York, 1964.
7. Brunfelt, A. O., Steinnes, E., "Determination of Selenium in Standard Rocks by Neutron Activation Analysis," *Geochim. Cosmochim. Acta* (1967) **31**, 283.
8. Wells, N., "Selenium Content of Soil-Forming Rocks," *N. Z. Geol. Geophys.* (1967) **10**, 198.
9. Rankama, K., Sahama, Th. G., "Geochemistry," University of Chicago Press, Chicago, Ill., 1950.
10. Palache, C., Berman, H., Frondel, C., "The System of Mineralogy of James Dwight Dana and Edward Salisbury Dana," Vol. 1, "Elements, Sulfides, Sulfo-salts, Oxides," 7th ed., Wiley, New York, 1944.
11. Brasted, R. C., "Comprehensive Inorganic Chemistry," Vol. 8, "Sulfur, Selenium, Tellurium, Polonium, and Oxygen," Van Nostrand, Princeton, N. J., 1961.
12. Suzuoki, T., "A Geochemical Study of Selenium in Volcanic Exhalation and Sulfur Deposits," *Bull. Chem. Soc. Jap.* (1964) **37**, 1200.

13. Latimer, W. M., "The Oxidation States of the Elements and Their Potentials in Aqueous Solutions," 2nd ed., Prentice-Hall, Englewood Cliffs, N. J., 1952.
14. Fleming, G. A., "Selenium in Irish Soils and Plants," *Soil Sci.* (1962) **94,** 28.
15. Byers, H. G., Williams, K. T., Lakin, H. W., "Selenium in Hawaii and its Probable Source in the United States," *Ind. Eng. Chem.* (1936) **28,** 821.
16. Byers, H. G., "Selenium Occurrence in Certain Soils in the United States, with a Discussion of Related Topics," *U. S. Dept. Agr. Tech. Bull.* (1935) **482,** 48.
17. Goering, H. R., Cary, E. E., Jones, L. H. P., Allaway, W. H., "Solubility and Redox Criteria for the Possible Forms of Selenium in Soils," *Soil Sci. Soc. Amer. Proc.* (1968) **32,** 35.
18. Gissel-Nielsen, G., Bisbjerg, B., "The Uptake of Applied Selenium by Agricultural Plants 2. The Utilization of Various Selenium Compounds," *Plant Soil* (1970) **32,** 382.
19. Vine, J. D., "Element Distribution in Some Paleozoic Black Shales and Associated Rocks," *U. S. Geol. Survey Bull.* (1969) **1214-G,** G1–G32.
20. Davidson, D. F., Lakin, H. W., "Metal Content of Some Black Shales of the Western Conterminous United States Part 2," *U. S. Geol. Survey Prof. Paper* (1962) **450-C,** C74.
21. Tourtelot, H. A., "Minor-Element Composition and Organic Carbon Content of Marine and Nonmarine Shales of Late Cretaceous Age in the Western Interior of the United States," *Geochem. Cosmochim. Acta* (1964) **28,** 1579.
22. Brimhall, W. H., "Progress Report on Selenium in the Manning Canyon Shale, Central Utah," Brigham Young University Geologic Studies, Vol. 10, pp. 104–120, 1963.
23. Tourtelot, H. A., Schultz, L. G., Gill, J. R., "Stratigraphic Variations in Mineralogy and Chemical Composition of the Pierre Shale in South Dakota and Adjacent Parts of North Dakota, Nebraska, Wyoming, and Montana," *U. S. Geol. Survey Prof. Paper* (1960) **400-B,** B447–B452.
24. Hashimoto, Y., Hwang, J. Y., Yanagisawa, S., "Possible Source of Atmospheric Pollution of Selenium," *Env. Sci. Technol.* (1970) **4,** 157.
25. Pillay, K. K. S., Thomas, C. C., Jr., Kaminski, J. W., "Neutron Activation Analysis of the Selenium Content of Fossil Fuels," *Nucl. Appl. Technol.* (1969) **7,** 478.
26. Shah, K. R., Filby, R. H., Haller, W. A., "Determination of Trace Elements in Petroleum by Neutron Activation Analysis II. Determination of Sc, Cr, Fe, Co, Ni, Zn, As, Se, Sb, Eu, Au, Hg, and U," *J. Radioanal. Chem.* (1970) **6,** 413.
27. Lansche, A. M., "Selenium and Tellurium—A Materials Survey," *U. S. Bur. Mines Inf. Circ.* (1967) **8340.**
28. Kubota, J., Allaway, W. H., Carter, D. L., Cary, E. E., Lazar, V. A., "Selenium in Crops in the United States in Relation to Selenium-Responsive Diseases of Animals," *J. Agr. Food Chem.* (1967) **15,** 448.
29. U. S. Department of Health, Education, and Welfare, "Public Health Service Drinking Water Standards," *Public Health Serv. Publ.* (1962) **956.**
30. Kharkar, D. P., Turekian, K. K., Bertine, K. K., "Stream Supply of Dissolved Silver, Molybdenum, Antimony, Selenium, Chromium, Cobalt, Rubidium and Cesium to the Ocean," *Geochim. Cosmochim. Acta* (1968) **32,** 285.
31. Lindberg, P., "Selenium Determination in Plant and Animal Material, and in Water. A Methodological Study," *Acta Vet. Scand., Suppl.* (1968) **23.**
32. Scott, R. C., Voegeli, P. T., Sr., "Radiochemical Analyses of Ground and Surface Water in Colorado, 1954–1961," Colorado Water Conservation Board Basic-Data Report No. 7, 1961.

33. Schutz, D. F., Turekian, K. K., "The Investigation of the Geographical and Vertical Distribution of Several Trace Elements in Sea Water Using Neutron Activation Analysis," *Geochim. Cosmochim. Acta* (1965) **29,** 259.
34. Chau, Y. K., Riley, J. P., "The Determination of Selenium in Sea Water Silicates and Marine Organisms," *Anal. Chim. Acta* (1965) **33,** 36.
35. Lakin, H. W., Byers, H. G., "Selenium Occurrence in Certain Soils in the United States, with a Discussion of Related Topics," Sixth Report, *U. S. Dept. Agr. Tech. Bull.* (1941) **783.**
36. Hashimoto, Y., Winchester, J. W., "Selenium in the Atmosphere," *Environ. Sci. Technol.* (1967) **1,** 338.
37. Weiss, H. V., Koide, M., Goldberg, E. D., "Selenium and Sulfur in a Greenland Ice Sheet: Relation to Fossil Fuel Combustion," *Science* (1971) **172,** 261.
38. Lewis, B. G., Johnson, C. M., Delwiche, C. C., "Release of Volatile Selenium Compounds by Plants. Collection Procedures and Preliminary Observations," *J. Agr. Food Chem.* (1966) **14,** 638.
39. Evans, C. S., Asher, C. J., Johnson, C. M., "Isolation of Dimethyl Diselenide and Other Volatile Selenium Compounds from *Astragalus racemosus* (Pursh.)," *Aust. J. Biol. Sci.* (1968) **21,** 13.
40. Shrift, A., "Biochemical Interrelations Between Selenium and Sulfur in Plants and Microorganisms," *Fed. Proc., Fed. Amer. Soc. Exp. Biol.* (1961) **20,** 695.
41. Cooper, W. C., "Selenium Toxicity in Man" in "Symposium: Selenium in Biomedicine," pp. 185–199, Avi, Westport, Conn., 1967.
42. Abu-Erreish, G. M., Whitehead, E. I., Olson, O. E., "Evolution of Volatile Selenium from Soils," *Soil Sci.* (1968) **106,** pp. 415–420.
43. Allaway, W. H., Kubota, J., Losee, F., Roth, M., "Selenium, Molybdenum, and Vanadium in Human Blood," *Arch. Environ. Health* (1968) **16,** 342.
44. Kifer, R. R., Payne, W. L., Ambrose, M. E., "Selenium Content of Fish Meals II," *Feedstuffs* (1969) **41,** 24.
45. Lunde, G., "Analysis of Arsenic and Selenium in Marine Raw Materials," *J. Sci. Food Ag.* (1970) **21,** 242.
46. "Analysis of Trace Elements in Seaweed," *J. Sci. Food Agi.* (1970) **21,** 416.

RECEIVED January 7, 1972. Publication authorized by the Director, U. S. Geoligical Survey.

7

Functional Aspects of Boron in Plants

W. M. DUGGER

Department of Biology, College of Biological and Agricultural Sciences, University of California, Riverside, Calif. 92502

Although the exact role of boron in plants is unknown, several physiological and biochemical activities associated with tissue boron content have been supported experimentally. This review covers some recent work on the role of boron in (1) organic translocation in plants, (2) enzymatic reactions, (3) plant growth regulator response, (4) cell division, (5) cell maturation, (6) nucleic acid metabolism, (7) phenolic acid biosynthesis, and (8) cell wall metabolism. The confusion surrounding the metabolic role of boron in plants has resulted in a "search" for the initial physiological effect of removing boron from the plant environment or, in some cases, adding boron to the plant growth medium or to excise tissue culture. Early plant responses to boron or to its deficiency and other types of experiments have helped clarify boron's role in plant metabolism.

Although the fact that boron is essential for higher plants has been known for 62 years, a specific single role has never been elucidated; instead a number of roles for boron in plant metabolism have been postulated (*1*). Within the past 50 years much research has been directed toward determining the element's physiological role. In recent years, it has become apparent that the "primary" role, if there is just one, may be associated with a biochemical effect of boron in enzymatic reactions or at the nucleic acid biosynthesis level. Much additional work will have to be done before the apparently separate and distinct roles can be brought together into a generally acceptable theory. This review attempts to consolidate some of the recent literature that points out diverse ideas on how boron acts in plant metabolism.

Organic Translocation in Plants

In 1953 a hypothesis was presented to account for one essential role of boron in plants (*2*). The authors interpreted their experimental data to mean that boron, by virtue of its complexing ability with sugars, facilitated the translocation of sugars in plants. Data were presented showing a significant increase in O_2 uptake by root tissue when exogenous substrate (sucrose) contained 5 ppm of boron. In addition, ^{14}C sugar supplied to intact leaves of bean and tomato plants entered and was distributed throughout the plant to a greater degree when 10 ppm of boron were present in the labeled solution. In subsequent work (*3*), it was shown that greater percentages of endogenous ^{14}C-labeled photosynthate, formed *in situ* by plants supplied with $^{14}CO_2$, were translocated by boron-sufficient plants than by plants grown on boron-deficient medium. Differences were seen as early as two days after boron was removed from the plants. At the same time, it was observed that earlier symptoms of boron deficiency could not be alleviated by sugar sprays applied to the plant.

Workers in Canada (*4*) found that ^{14}C sugars applied to primary leaves of bean plants with a wetting agent were distributed more widely throughout the leaves when boron was added to the mix. They observed that lowering the pH of the labeled solution had no effect on reducing the uptake of ^{14}C sugars and concluded that the effect of boron was not caused by its sugar-complexing ability as previously suggested (*2*). Skok (*5*) also concluded that it did not appear that boron was functional in sugar translocation because of its complexing ability, and Scholz (*6*) reported that boron was transported by the plant exclusively in one direction—from the roots to the shoot. His work did not support the sugar–borate complex transport hypothesis. In a review (*7*), Skok suggested that boron's capacity to complex with polyhydroxy compounds appears to be related to its physiological function but that the nature of the relationship was not clear. He further suggested that the apparent effect of boron on sugar translocation was indirect.

Another method demonstrating boron's effect on sugar translocation was reported by Mitchell *et al.* (*8*). Previous work had shown that the translocation of plant hormones such as (2,4-dichlorophenoxy)acetic acid (2,4-D), IAA, and 1-naphthaleneacetic acid (NAA) from leaves to other plant parts was associated with translocation of photosynthate. When boric acid was added to the growth regulator applied to primary leaf of bean plants, the stem curvature induced by the unilateral application of growth regulator was accelerated. The addition of sugar to the hormone–borate–wetting agent mix resulted in an additional acceleration of hypocotyl curvature. Measurable differences in curvature of the hypocotyl occurred 1.5 hr after the hormone–sugar–borate mixture, was applied.

Dyar and Webb (*9*) observed that significantly less ^{14}C-labeled photosynthate was translocated to the vegetative terminal region in boron-deficient bean plants than in boron-sufficient plants. However, they reported that NAA applied to the terminal bud region of boron-deficient plants counteracted the boron deficiency with respect to the movement of ^{14}C label into the vegetative buds. They suggested that boron performs an essential role in the biosynthesis of auxins within the plant meristem and that the increased translocation of sugars occurs as a result of auxin-stimulated growth rather than the reverse. These investigators did not report any results in which IAA, the natural auxin in plants, overcame boron deficiency.

Perkins (*10*), as part of an extensive study on the chemistry of boron in plants, reported that the uptake of ^{14}C-fructose, supplied exogenously to the petiole stump of excised trifoliate leaves of soybean plants, was markedly increased when 10 or 50 ppm of boron were supplied with the fructose. At comparable sections down the stem of the plants, the label was greater in the boron-supplied plants. However, the author stated that no conclusion about the effect of boron on increased translocation in plants was justified from his data. He speculated, however, that boron, "with reasonable certainty," increased the uptake of sugars by plant cells but played only an indirect role in sugar translocation.

Turnowska-Starck (*11*) reported that plants grown under a low level of boron nutrition, but without deficiency, and fed ^{14}C-sucrose through the leaves, exhibited a much lower level of radioactivity than plants supplied with a higher level of boron in the medium. When boron was included with the ^{14}C sugar, there was an increased absorption of sugar but not necessarily more transport down the stem. Others (*12*) have also concluded that boron enhances the migration of foliar-applied sugars to other plant parts, but these investigators do not believe that such facilitated uptake into the leaves provides evidence that boron stimulates translocation *per se*. Their experimental procedure for studying translocation involved a long period of labeling of entire leaves with $^{14}CO_2$.

On the other hand, Liang and Tsao (*13*) observed that boron treatment of cotton plant leaves increased sugar in the boll opposite treated leaves. Saakov (*14*) also found that in boron-sufficient bean plants ^{14}C-glucose applied to the cut petiole moved into other plant parts more rapidly than in plants exhibiting boron deficiency (12–15 days after removal of boron from the nutrient medium). It was also observed that the distribution of ^{14}C down the stem of boron-deficient sunflower plants after 30–60 min in $^{14}CO_2$ was less than in control plants with boron (*15*). Although the transport profile was the same, the advancing front of radioactivity in the boron-sufficient plants was consistently further down the stem within a given time interval than in boron-deficient plants. The authors concluded that the velocity of translocation was reduced in the boron-deficient plants. The difference between the amount of ^{14}C entering the deficient and control plants was not significant; however, the amount of label entering the translocation system in boron-deficient plants was less. Nelyubora and Dorozhkina (*16, 17*) also reported on the effect of boron nutrition on ^{14}C-sugar movement in plants, using carrots and sugar beets in their study. Boron-deficient plants showed an inhibited transport of labeled sugar to vegetative portions of the plants as well as to storage roots.

Skok (*7*) states that it appears that there is "some relationship between boron and sugar translocation . . . ," and since boron–sugar complexes have not been separated from the leaves of translocated systems of plants, perhaps the relationship is "indirect" and involved with other cellular activities more directly influenced by cellular boron content. However, a substantial number of data do indicate an early and possibly direct effect on both sugar movement (*2, 3*) and sugar-dependent growth regulator movement from bean leaves to hypocotyl (*8*). In the latter case, considerable difference was observed 2–3 hr after application of a borate–sugar–growth regulator mix to the leaves of soil-grown, boron-sufficient plants. These observed changes were much too fast to be dependent on growth or developmental differences in plants sufficient enough in boron content to maintain maximum growth rate.

Enzymatic Reaction in Plants

In two early reviews (*1, 7*), boron as it affects plant enzymes was briefly discussed. One study involved the influence of boron on potato aldehyde oxidase, tyrosinase, and starch phosphorylase (*18*). No effect was observed on the former two enzymes with *in vivo* studies; however, a 21% inhibition in starch synthesis from glucose 1-phosphate by a 0.1*M* concentration of boron was reported. Reed (*19*) reported that boron deficiency in celery plants resulted in increased activity of catechol oxidase. The oxidation product, quinones, polymerized to phenolic or melanitic aggregates which typified boron deficiency in plants. Boron-deficient tissue was also shown by MacVicar and Burris (*20*) to have a more active polyphenol oxidase than that present in normal tissue. Roush and Norris (*21*) observed that boron was a competitive inhibitor of xanthine oxidase. The inhibition could be reversed by sorbitol and less effectively by glucose or ribose. Later it was reported that borate inhibited the oxidation of tyrosinase (*22*), and Roush and Gowdy (*23*) observed the competitive inhibition of alcohol dehydrogenase by borate. The inhibition was prevented by including ribose, sorbitol, or manitol in the reaction mixture. They suggested two possible mechanisms of borate inhibition: (a) competition of a borate–NAD complex with free NAD for enzyme surface, which was supported by the comparable magnitude of K_m (NAD) and K_i (borate), and (b) borate binding to the zinc ion of the enzyme or at the positive nitrogen moiety of the pyridine. The high ratio of ribose to NAD required to prevent borate inhibition suggests this possibility. Borate was also observed inhibiting the alkaline phosphotase activity in milk and intestinal mucosa (*24*).

In an interesting study, Alvarado and Sols (*25*) used borate to complex with fructose 6-phosphate, the product of phosphomannose isomerase, thereby inhibiting phosphoglucose isomerase and making it possible to assay the activity of phosphomannose isomerase. It was also noted that boron inhibited the *in vitro* conversion of glucose 1-phosphate to starch phosphorylase (*26*). The authors proposed that because boron inhibited starch synthesis, the earlier observed influence of boron on sugar translocation (*2, 3*) may be a function of the higher level of sugars in leaf cells at any one time. In boron-deficient plants, the enzyme was not inhibited, and, therefore, starch accumulated. Scott (*27*) confirmed the effect of boron on starch phosphorylase and proposed that "boron performs a protective function in plants, in that it prevents excessive polymerization of sugars at sites of sugar synthesis (it seems significant that the highest concentration of boron in the plant is in the leaves where sugars are actively synthesized)." Scott extended the hypothesis to include the role of boron in sugar polymerization.

Other researchers observed that boron stimulated uridine diphosphate glucose (UDPG) pyrophosphorylase but inhibited the reaction catalyzed by UDPG transglycosylase (*28*). However, with sugar cane leaf and pea seedling homogenates, sucrose synthesis was enhanced when the reaction mixture contained boron. Although these investigators did not observe the borate inhibition of phosphoglucomutase from muscle, Loughman (*29*) showed that this enzyme in pea seeds was inhibited by boron. He suggested that if starch utilization is blocked by borate's effects at the phosphoglucomutase stage, two consequences are possible: (a) glucose 6-phosphate, fructose 6-phosphate, and fructose formation would be inhibited, thus reducing the amount of fructose 6-phosphate and fructose available for sucrose synthesis; and (b) more glucose 1-phosphate would be available for UDPG synthesis, and, with fructose limited because of phosphoglucomutase inhibition, UDPG would be available for other cellular reactions, such as galacturonic acid synthesis or other cell wall constituents. Loughman (*29*) suggested that the mutase enzyme from pea seeds was different from the muscle enzymes. The primary boron effect was the result of a specific effect on the plant protein and not one on the substrate complexing ability.

Maevskaya and Alekseeva (*30*) reported a marked reduction in the ATP content of young sunflower plant terminal buds grown under boron deficiency. The level of ATPase in such plants was about twice that in boron-sufficient plants. Hinde *et al.* (*31*), also observed that the ATPase in boron-deficient bean seedling radicals was greater than in control plants. Boron-deficient roots showed an increase in acid phosphatase over the control and a decrease in alkaline pyrophosphatase and amino acid dependent ATP pyrophosphatase exchange (*32*). The author suggested that for all four enzymes the same type of activity change induced by boron deficiency occurs with normal roots in sections removed from the tips. It appears likely that boron-deficiency effects may be a general morphological and physiological change remote from the original cellular event that is susceptible to boron deficiency.

Shkol'nik (*33*) showed that boron deficiency in plants reduced the RNA and DNA content in the meristem and that RNAase activity was increased. A similar observation was made by Arbal (*34*). In a later paper (*35*), it was reported that addition of 8-azaguanine to boron-sufficient plants also increased RNAase activity. The investigators suggested a similarity in morphological changes in plants caused by boron deficiency and 8-azaguanine poisoning. Timashov (*36*) observed an increased breakdown of nucleotides by cellular organelles from boron-deficient sunflower plants.

A significant study on the effect of boron on plant metabolism was made by Lee and Aronoff (*37*). They demonstrated that boron complexed with 6-phosphogluconate, thereby inhibiting 6-phosphogluconate dehydrogenase. Without borate in the reaction mixture, the enzyme inhibition was released and substrate was metabolized to a greater extent through the pentose phosphate shunt. Such pathway changes lead to the synthesis of phenols, which in turn might further complex with borate, decreasing still more the available boron for modulating the 6-phosphogluconate dehydrogenase. With this further increase in metabolic activity of the pentose phosphate shunt in boron-deficient plants, an autocatalytic formation of phenolic acids occurs. A build-up of phenolic acids does occur in boron-deficient plants (*7*).

Parish (*38*) studied the *in vitro* effect of boron on peroxidase from horseradish. Aerobic oxidation of IAA by the enzyme was slightly stimulated by boron, and the lag phase induced by chlorogenic acid (a plant product under boron deficiency) was reduced by the element. Boron also seemed capable of protecting IAA from peroxidasic action under anaerobic conditions. Binding between the enzyme and boric acid did occur, but the prosthetic heme group did not seem to be involved. Rather, the author seemed to think that the complexing property of the enzyme with boron was the function of the sugar moiety. In an *in vivo* study, Parish (*39*) reported that peroxidase activity in *Vicia faba* seedlings was greater in boron-deficient tissue. Odhnoff (*40*) also reported that root tips of boron-deficient bean plants showed an increased peroxidase activity. These reports are in contrast to an earlier study where it was observed that sunflower stems or root tissue culture grown without boron had less peroxidase activity (*41*). Parish (*39*) also observed that boron may facilitate the attachment of peroxidase enzyme to the cell wall. In contrast, calcium facilitates the release of enzyme from the wall.

Plant Growth Regulator Response

Reference was made earlier to the possible interrelationship of boron and plant hormones synthesis and activity (*1*). This interrelationship has been the subject of several recent research reports that make significant contributions to the understanding of boron action in plants. Tomato and turnip plants grown for three weeks under boron-deficient conditions and sprayed with sugar and boron grew better and were more like control plants than unsprayed plants (*42*). However, Odhnoff (*43*) concluded that there was no relationship between boron and IAA in growth tests when the hormone and element were added in various concentrations to the nutrient solution. She presumed that boric acid had a growth regulator effect, probably in an indirect manner, influencing an enzyme system such as peroxidase. Perkins (*10*) studied the effect of boron on oxidation of several phenolic compounds by manganic ions [Mn(III)]. At pH 8 the presence of 10 and 100 ppm of boron markedly inhibited the oxidation of phenols. His thesis was that phenols complex in the presence of boron, thereby reducing the substrate level for the phenol oxidase–peroxidase enzyme system. This, in turn, would inhibit the rate of manganic ion formation from manganous ions [Mn(II)], thereby inhibiting the oxidation of IAA. If this mechanism were operating, then boron-deficient plants would be deficient in IAA. No data were presented to confirm this hypothesis.

Dyar and Webb (*9*) presented an interesting hypothesis as a result of their study of the relationship between boron and NAA as it influences translocation of ^{14}C-labeled endogenous material in bean plants. They theorized "that boron plays an essential role in the biosynthesis of auxins in the meristems of the plant, translocation occurring as a result of growth rather than the reverse."

Boron-deficient sunflower plants contained more IAA than did normal plants (*44*). This difference was apparent at the pre- and post-deficient stages of growth. Although the investigators did not present data to show that boron-deficient plants had an inhibited IAA oxidase, they did propose that in the absence of boron the phenolic compounds that accumulate under such conditions inhibit IAA oxidase. Consequently, IAA in deficient tissue would be at a greater than normal level. They further proposed that boron controls the level of IAA in plants, keeping it within physiological growth-promoting concentrations.

Coke and Whittington (*45*) experimentally explored the hypothesis that boron deficiency is equivalent to IAA toxicity. Extracts of roots from boron-deficient plants were much more inhibitory to growth of bean root segments than extracts from normal roots. When roots were grown in excess IAA, they recovered more quickly if boron were present in high concentrations. These investigators suggested that boron deficiency induced excess auxin may be caused by one of two possibilities: (a) the build-up of phenolic inhibitors which inhibit the IAA oxidation system (normally boron inactivates such inhibitors by complexing with them), or (b) some fundamental growth process such as cell-wall synthesis or nucleic acid synthesis which is impaired by boron deficiency. Therefore, IAA is not used in the growth or metabolic process and accumulates to an excess in cells. As pointed out in the previous section (*38*), boron is capable of modifying reactions involving peroxidase and IAA. Saini *et al.* (*46*) also proposed that boron aided in the initiation of cotton plant flowers by suppressing IAA activity.

Nucleic Acid Biosynthesis

In studying the effect of boron on the growth of bean radicals, Whittington (*47*) observed that it was unlikely that cell division ceased because of a shortage of DNA. In contrast, he observed that there was more DNA in boron-deficient roots and suggested that the higher level might be caused by the physiologically older root cells near the tip in the boron-deficient plants. This study revealed that the RNA content per cell in boron-deficient root tissue was less than in controls.

In an extensive series of reports, Shkol'nik and colleagues investigated the role of boron in nucleic acid content of boron-deficient tissue as compared with control tissue and found that it was possible to alleviate boron deficiency symptoms to some degree by supplying nucleic acid to the growing media (*49*). The incorporation of ^{32}P into nucleic acid of sunflower was strongly inhibited by boron deficiency (*50*). In addition to influencing the level of nucleic acid content, boron-deficient plants had an increased activity of RNAase and boron-sufficient tissue incorporated ^{14}C-adenine into nucleic acid and labeled tyrosine into protein more than boron-deficient tissue (*33, 51*). Sherstnev and Kurilenok (*52*) had reported earlier on the boron stimulation of ^{14}C-adenine incorporation into RNA of young leaves and roots of sunflower plants. It was also observed that the amount of several amino acids was higher in plants grown on boron-deficient medium than in control plants whereas the incorporation of ^{14}C-tyrosine into protein was several times less (*53*).

A similar observation was made with ribosomal fractions from boron-sufficient and boron-deficient sunflower plants (*54*). The disturbed activity of polyribosomes in boron-deficient tissue resulted from the inhibition of RNA synthesis and an increase in RNAase activity (*55*). Timashov (*36*) also observed that the exclusion of boron from the nutrient solution increased the breakdown of nucleotides by chloroplast, mitochondria, and supernate fraction from young sunflower leaves and cotyledones. The addition of an inhibitor, heparin, to the nutrient solution without boron temporarily aided in alleviating boron-deficient symptoms (*56*). Total nitrogen was observed to be high in boron-deficient peanut plants, and there was an increase in the level of arginine, aspartic acid, glutamic acid, proline, and serine. On a plant cell basis (milligrams of DNA), protein remained the same as control plants. The DNA content per gram of dry weight also remained the same for deficient and sufficient leaves (*57*).

Borshchenko (*58*) observed that boron-deficient pea roots produced normal amounts of aminoacyl tRNA but incorporated less amino acid into protein. No polysomes were found in the boron-deficient roots. Albert (*59*) also observed that in roots of boron-deficient tomato plants the RNA content decreased within 24–48 hr after boron was removed from the growth medium. However, root elongation ceased within 6–12 hr before there was a noticeable decrease in RNA. Later it was observed that the addition of purine, guanines, and cytocine to the nutrient solution stimulated root elongation when boron was not present (*60*). When boron was supplied to seedlings that were beginning to show deficiency symptoms, Hundt *et al.* (*61*) observed an increased incorporation of ^{32}P into RNA and DNA in young leaves and roots before an increase in net protein synthesis occurred. Rapota (*62*) reported that pea seedlings grown in a boron-free medium incorporate less ^{3}H-uridine into RNA and ^{35}S-methionine into protein than plants supplied boron. The measurement of DNA synthesis by the incorporation of ^{3}H-thymine revealed that DNA synthesis was eliminated by the deficiency. In contrast to the previous report, Cory *et al.* (*63*) found that there was increased incorporation of labeled precursors into the nucleic acid of radicals in boron-deficient bean plant. The difference was observed within 4 hr after placing the seedlings in a deficient medium. The specific activity of soluble RNA, DNA, and ribosomal RNA was greater in boron-deficient than in normal tissue. This effect was judged to be caused by an increasing incorporation of nucleotide into the nucleic acids (*64*).

Cellular Growth and Differentiation

A rather complete review of boron as it relates to cellular maturation and differentiation, cell wall formation, and pectic synthesis was made by Skok (*7*). Recently Shkol'nik (*51*) reviewed in detail the role of boron in a multitude of cellular processes and attempted to form a unifying theory on the essentiality of boron in plants. Presently, as earlier, theory holds that boron does function in some way in cellular growth and maturation. The exact mechanism is unknown; however, some research has been conducted recently on this role of boron. Odhnoff (*40*) investigated boron deficiency in bean plants as it influences growth. She found an increase in cell wall materials of plant parts showing deficiency symptoms. Since root growth was impaired before transport, she interpreted her data as demonstrating that the primary influence of boron was on the stretching phase of cell elongation (*43*).

In a series of reports, Whittington and co-workers investigated the role of boron in plant growth. The first study (*65*) reported that cell division in bean plants ceases rapidly under boron deficiency. This lack of meristem growth maintenance was reported to be responsible for abnormalities associated with boron deficiency. In a second study (*47*), Whittington conducted an extensive investigation of cell number, cell volume, and organic compound analysis on 1-mm root sections of bean radicals from seedlings grown under boron-deficient and boron-sufficient nutrient conditions. He reiterated his thesis that the primary effect of boron deficiency was on cell division as well as cell enlargement of the apical cells. He also explored the hypothesis that boron's role in cell division and elongation may be concerned with pectin synthesis in the division and elongation stages. The third report in this series (*66*) showed a decreased incorporation of ^{14}C-glucose into pectin by root sections, with an increased incorporation into cellulose. These results suggested that boron was involved in cell wall bonding and was not required for specific enzymes involved in cell wall synthesis.

Skok (*67*) found that when boron was withheld from sunflower seedlings, the plants became more resistant to x-irradiation. He suggested that cellular maturation and not cell division was the phase of growth interfered with by boron deficiency.

In an investigation of boron requirement for flax roots in sterile culture, Neales (*68*) found that the effect of boron on lateral root growth was consistent with the effect the element has on meristematic growth in intact plants. He supplied ^{14}C-sucrose to the cultures for 12 hr and determined the label in glucose, fructose, and sucrose in the extracts from normal and boron-deficient roots. The conclusion was that root growth was not restricted by the lack of sugar movement into root cells. With tomato root cultures, Neales associated boron requirements with cell division or cell expansion (*69*). In a later study (*70*), no evidence was found to associate boron with calcium metabolism in bean radical growth. By substituting organoboron compounds for boric acid in a study of bean plant root growth, Wildes and Neales (*71*) concluded that their results support "the hypothesis that the activity of borate as an essential nutrient, depends upon its ability to form a biologically active complex with an *in vivo* cis-diol compound."

Albert and Wilson (72) investigated the effect of boron deficiency on elongation of tomato plant roots. Cessation of root elongation was noticeable within 6 hr after boron was withheld. The first internal disorder of deficiency became apparent within the first 24 hr after plants were transferred to boron-free medium. The first symptom appeared to be disintegration of the protoplast of some cortical cells. Root elongation in cultures with an adequate supply of boron correlated positively with the total solar radiation of the two days before the root elongation measurement period. With boron deficiency, elongation was negatively correlated with total solar radiation for the first day of the measurement period. Humphries (73) observed that boron had little effect on root formation in plants illuminated with fluorescent lights although with incandescent light, root formation depended on the boron supply. It was also observed by MacInnes and Albert (74) that the boron-deficient symptoms of tomato (decreased root elongation, increased coloration and decreased RNA content) develop more rapidly at high than at low light intensities.

Yih and Clarke (75) investigated the effect of boron on root elongation and carbohydrate and nitrogen content of tomato plants grown in solution cultures. The level of protein and carbohydrate content in the root tips of boron-deficient plants was greater than that in control plants, and the authors suggested that this reflects the effect of boron deficiency on enhanced maturation of root cells closer to the root tip. The observed increase in root laterals close to the tip in deficient plants is related to the increased maturation of cells near the meristem. After 72 hr of growth, under sufficient and deficient boron conditions, the roots were longer by a factor of 6.5 in shade-grown plants and by a factor of 20 in unshaded grown plants where boron was in the media. If one assumes that the carbohydrate level in the terminal 10-mm root sections was consistent through the root length, the product of growth times total sugar content was 3.5 more in roots of the shaded plus boron plants and 21 more in the unshaded plus boron plants than in the corresponding deficient plants. Perhaps the differences in levels of sugar (about 2 times in the deficient roots) do not truly account for the carbohydrate utilized in the considerably greater root growth of the control plants. In effect, the total carbohydrate movement to the roots of sufficient plants was evidently much more than in the deficient plants, with much of it being used for continuing root growth. However, the investigators did not see any correlation between the elongation and the carbohydrate or protein nitrogen content of the root tips.

Albert (*59*) also found that elongation of tomato plant roots ceased before there was a decrease in RNA content (*see* previous section). Examination of leaf tissue from boron-deficient plants with the aid of the electron microscope shows that fine cellular structure changes occur soon after removing boron from the nutrient solution. Chloroplasts degenerate, starch accumulates, cell walls have a different structure, mitochondria increase in number with myline figures, and nuclei develop dense rhombohedral structures before the typical deficient symptoms become apparent (*76*).

An interesting observation on the inheritance and physiology of boron response in tomato plants was observed by Wall and Andrus (*77*). An abnormal phenotype caused by boron deficiency was found to be controlled by a single recessive gene. Plants of the genetic variant contained considerably less boron in their leaves and more in their roots than the controlled variety when grown in high boron-containing nutrient solutions. A similar variant in red beet (*Beta vulgaris*) was reported by Tehrani *et al.* (*78*).

Phenolic Acid Biosynthesis and Lignification

Reed (*19*) and others cited by Skok (*7*) reported that boron deficiency generally leads to an accumulation of phenolic compounds. This has been thought to result in a decreased level of lignins in plants—a product of phenolic polymerization in boron-sufficient tissues. The boron-deficient induced fluorescence in celery tissue observed by Spurr (*79*) was believed to be caused by the accumulation of caffeic and chlorogenic acids in such tissue, a possible secondary effect of boron deficiency (*80*). It was further suggested (*81*) that the necrosis of tissue caused by boron deficiency arises from an increase in caffeic acid. This acid or some product resulting from its further metabolism in boron-deficient tissue was proposed to cause destruction of plant conductive tissue. Perkins (*10*) observed that ^{14}C-shikimic acid fed to the boron-deficient tissue incorporated less of the labeled compound in the lignin and/or melanin pigment fractions (90% ethanol-insoluble fraction): "The lignin synthesizing ability of boron deficient sunflower leaves apparently decrease with increasing severity of the symptoms."

Watanabe *et al.* (*82*) also observed the accumulation of scopoletin glucoside in boron-deficient tobacco leaves. The leaves from boron-deficient plants contained a 20-fold increase in the glucoside compared with normal tobacco plants. Other blue fluorescent compounds accumulating in boron-deficient tissue have also been identified (*83*). Boron-deficient culture medium used for sunflower stem callus and root tissue culture resulted in less lignification than in normal tissue (*41*). Recently,

Rajaratnam *et al.* (*84*) reported that boron-deficient oil palms showed a total absence of the flavonoid, leucoanthocyanines, before visible symptoms were apparent. Corn grown under boron deficiency also showed fewer flavonoids in the leaves during growth (*85*).

Carbohydrate Metabolism and Respiration

It was assumed that the observed increase in respiration when boron was added to the respiratory vessel containing bean root tips and sucrose was caused by the facilitated increase in sucrose uptake by the tissue (*2*). Boron-deficient cotton and turnip plants were observed to have a lower percent total sugar and a higher percent starch content than control plants (*42*). Considerably higher levels of respiration in bean leaf tissue also were observed when the tissue was infiltrated with boron and 4% glucose than with glucose alone (*26*). Neales (*86*) found that boron deficiency in flax plants caused accumulation of more sugar in the stems than occurred in normal plants, and Odhnoff (*40*) observed that boron-deficient plants contained a significant increase in reducing as well as nonreducing sugars in roots and leaves compared with normal plants. Scott (*28*) also observed that respiration of sunflower leaves in plants treated with a super optimum level of boron was higher as was the concentration of total sugars plus starch. However, the starch content alone was lower in treated leaves than in the controls. Scholz (*87*) observed that boron deficiency in *Lemma minor* resulted in an extreme accumulation of starch, primarily as a result of the effect of boron on carbohydrate metabolism rather than any influence on translocation. Liang and Tsao (*13*) observed that cotton leaf tissue and bolls had increased starch and reducing sugar content when the opposite leaves were treated with boron, and as previously cited, boron-deficient tomato roots contained more carbohydrate than boron-sufficient roots (*75*). Differences in carbohydrate as well as protein appeared to reflect the influence of boron deficiency on maturation of tissue closer to the root tip and the promotion of lateral roots. Kull (*88*) found that the application of phenylboric acid resulted in an increased content of storage carbohydrate whenever there was an inhibition in longitudinal growth. There was no direct indication of the influence of phenylboric acid on carbohydrate metabolism. The respiration of sunflower leaves and terminal buds decreased as boron deficiency developed (*89*). Prior to the development of boron-deficient symptoms, there was an increase in oxygen uptake. Under boron deficiency, there appeared to be some alteration in normal metabolism, with a possibility that the respiration ratio from the glycolysis–pentose phosphate shunt was altered in boron-deficient tissue. Timashov (*90*) also observed an increase in the proportion of substrate metabolized by the glucose monophosphate shunt in boron-deficient sunflower tissue. Respiration in the boron-deficient tissue was also more resistant to KCN action than respiration in normal tissue.

Cell Wall and Membrane Metabolism

If boron plays a regulatory role in hexose sugar metabolism, it might be expected that there would be some influence of the element on cell wall metabolism, including pectin biosynthesis. Spurr (*91, 92*) did find that boron deficiency in celery plants altered plant cell walls. Collenchyma cell walls were thinner, and phloem parenchyma and ground parenchyma cell walls were thicker. There were fewer observed lamellae in the collenchyma cells, and some cells failed to develop typical "corner" thickenings. Spurr concluded that boron apparently affected the rate and process of carbohydrate condensation into wall material. Wilson (*93*) studied the effect of boron on the cell wall of tobacco pith parenchyma in tissue culture. The predominant effect of boron deficiency was a doubling of the amount of the cell wall fraction expressed in either dry or wet weight. There was little effect on the proportion of cellulose and pectic substances but a large decrease in galactan content, with some increase in arabans and xylans. Boron enhanced the incorporation of ^{3}*H-myo*-inositol into D-galacturonosyl and L-arabinosyl of pectic fractions from pollen tube membranes (*94*). The evidence suggests that boron plays a definite role in pectic synthesis of germinating pollen and the role may be related to the synthesis of D-galacturonosyl units.

Yih *et al.* (*95*) studied the boron requirement of pollen-derived tissue from *Ginko biloba*. They found that lower boron levels in the culture solution caused a reduced rate of cell division with no noticeable effect on cell size. The deficient boron level also resulted in a lower level of fructose in the hydrolizable polysaccharide fraction. They observed 23% more cell wall material in boron-deficient tissue. It has also been observed that potato tissue grown with boron sprayed on the foliage contained more phospholipids (*96*). Shkol'nik and Kopman (*97*) found that the root and apical parts of boron-deficient sunflower plants contained less phospholipids than normal plants. The authors attributed this decrease in phospholipid content to the role boron plays in altering the structural organization of cells.

Miscellaneous Effects

Many reports have been published concerning the role of boron in pollen germination (*1*). O'Kelley (*98*) observed that the oxygen uptake and sucrose or glucose absorption by germinating pollen were stimulated by boron. However, he believed that the effect of boron on pollen tube elongation was quite different and not related to the effect on either sugar absorption or respiration. Vasil (*99*) found that boric acid not only had a marked effect on the growth of the pollen tube, but also improved germination. Kumar and Hecht (*100*) found that in self-incompatible *Oenotheria organesis* the addition of boric acid and calcium to the water-treated styles enhanced pollen tube growth. The utilization of the stylus sugar was also stimulated by the addition of boric acid.

A higher concentration of pectins and possibly pentosans occurred in leaves of boron-deficient plants (*101*). This was interpreted to be a factor in the reduced wilting of boron-deficient plants under moisture stress. Leaves of boron-deficient plants also had a higher percentage of nonfunctional stomates and a higher sugar and colloid content. In addition, Starck (*102*) observed that without boron in the nutrient medium, plants absorbed less water and nutrient, possibly because of an indirect effect related to boron deficiency on root growth.

In general, with the exception of the diatoms (*103–105*) and perhaps blue-green algae (*106*), single-cell algae do not appear to require boron for continuous growth (*107–109*), nor do fungi generally require boron for growth (*51, 110*); pteridophytes, however, do appear to have a requirement for the element (*111*).

Prospectus

The attempt of Shkol'nik (*51*) to collate the various effects of boron deficiency on plant processes into a unifying theory could be a springboard for future research. He proposed that the primary alteration in metabolic processes when boron is withheld from the growing plant involves membrane breakdown. This breakdown, in turn, results in a release of RNAase from the bound, inactive form, which is followed by an alteration in nucleic acid and protein synthesis. The author suggested that these changes are related to the observed reduction in cellular phospholipids, membrane degeneration, increase in RNAase activity, and a possible shift in the catabolism of carbohydrates, with a larger fraction being oxidized *via* the pentose phosphate pathway. This in turn results in both an increased level of phenolic compounds and the inhibition of IAA oxidase. Consequently, the IAA level will be higher in boron-deficient plants. New techniques and methods in biochemistry and cell biology could be used to test various aspects of this hypothesis.

Several research reports on the role of boron in carbohydrate metabolism involving *in vivo* and *in vitro* studies can be brought together into a schematic diagram to illustrate additional need for clarifying its role in regulating synthesis and utilization of sugars and starch.

One possible scheme is illustrated below. The regulation of enzymatic reactions that result in a build-up of specific products (UDPG, glucose 1-phosphate, 6 phosphogluconate or others) may have significant effects in changing the physiology and/or morphology of plants. Lee

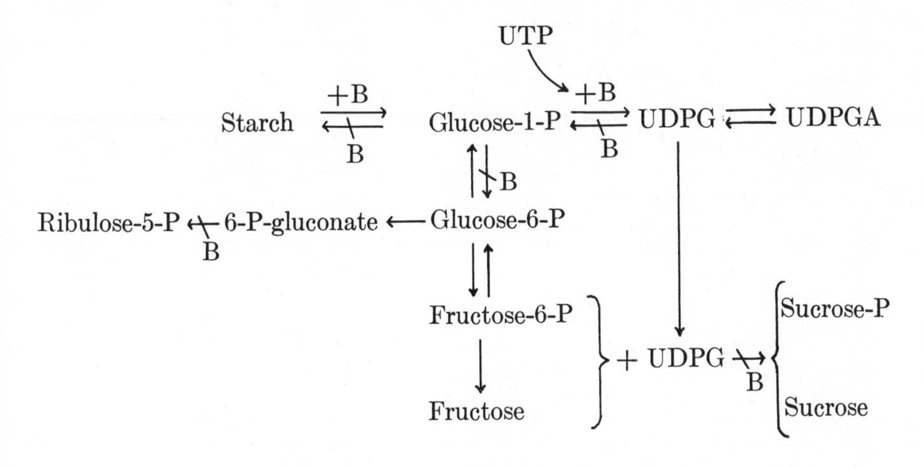

B in vitro reaction inhibited by boron

+B in vitro reaction stimulated by boron

and Arnoff (*37*) have pointed out the regulation of phenols by blocking 6-phosphogluconate dehydrogenase, Loughman (*29*) has shown that phosphoglucomutase is inhibited by boron, and it has also been reported (*26, 27*) that boron inhibits the synthesis of starch from glucose 1-phosphate and stimulates the synthesis of UDPG (*28*) a precursor for several diverse reactions leading to hexose elaboration into cellular constituents or products.

Additional effort to test this regulating role of boron in plant carbohydrate metabolism would be profitable.

Literature Cited

1. Gauch, H. G., Dugger, W. M., *Md. Agr. Exp. Sta. Bull.* (1954) **A-80.**
2. Gauch, H. G., Dugger, W. M., *Plant Physiol.* (1953) **28,** 457.
3. Sisler, E. C., *et al., Plant Physiol.* (1956) **31,** 11.
4. Nelson, C. D., Gorham, P. R., *Can. J. Bot.* (1957) **35,** 339.
5. Skok, J., *Plant Physiol.* (1957) **32,** 308.
6. Scholz, G., *Flora* (1960) **148,** 484.

7. Skok, J., in "Trace Elements," Chapter 15, p. 227, Academic, New York, N. Y., 1958.
8. Mitchell, J. W., *et al., Science* (1953) **118,** 354.
9. Dyar, J. J., Webb, K. L., *Plant Physiol.* (1961) **36,** 674.
10. Perkins, H. J., Ph.D. thesis, Iowa State College, 1957.
11. Turnowska-Starck, Z., *Acta Soc. Bot. Polen.* (1960) **29,** 533.
12. Weiser, C. J., *et al., Physiol. Plant.* (1964) **17,** 589.
13. Liang, Y. F., Tsao, T. H., *Acta Bot. Sinica* (1963) **11,** 178.
14. Saakov, V. S., *Dokl. Bot. Sci.* (1965) **162,** 89.
15. Lee, K., *et al., Physiol. Plant.* (1966) **19,** 919.
16. Nelyubora, G. L., Dorozhkina, L. A., *Izv. Tim. Sel. Akad.* (1968) **5,** 144.
17. Nelyubora, G. L., Dorozhkina, L. A., *Izv. Tim. Sel. Akad.* (1970) **1,** 118.
18. Winfield, M. E., *Aust. J. Exp. Biol. Med. Sci.* (1945) **23,** 267.
19. Reed, H. S., *Hilgardia* (1947) **17,** 377.
20. MacVicar, R., Burris, R. H., *Arch. Biochem.* (1948) **17,** 31.
21. Roush, A., Norris, E. R., *Arch. Biochem.* (1950) **29,** 344.
22. Yasunobu, K. T., Norris, E. R., *J. Biol. Chem.* (1957) **227,** 473.
23. Roush, A. H., Gowdy, B. B., *Biochem. Biophys. Acta* (1961) **52,** 200.
24. Zittle, C. A., Della Monica, E. S., *Arch. Biochem.* (1950) **26,** 112.
25. Alvarado, F., Sols, A., *Biochem. Biophys. Acta* (1957) **25,** 75.
26. Dugger, W. M., *et al., Plant Physiol.* (1957) **32,** 364.
27. Scott, E. G., *Plant Physiol.* (1960) **35,** 653.
28. Dugger, W. M., Humphreys, T. E., *Plant Physiol.* (1960) **35,** 523.
29. Loughman, B. C., *Nature (London)* (1961) **191,** 1399.
30. Maevskaya, A. N., Alekseeva, Kh. A., *Dokl. Bot. Sci.* (1965) **156,** 59.
31. Hinde, R. W., *et al., Phytochem.* (1966) **5,** 609.
32. Hinde, R. W., Finch, L. R., *Phytochem.* (1966) **5,** 619.
33. Shkol'nik, M. Y., *Int. Congr. Soil Sci.* (1967) **4,** 1215.
34. Arbal, Y. P., *Ind. J. Biochem.* (1966) **3,** 263.
35. Shkol'nik, M. Y., Smirnov, Y. S., *Tr. Bot. Inst. Akad. Nauk SSSR* (1970) **20,** 45.
36. Timashov, N. D., *Sov. Plant Physiol.* (1968) **15,** 778.
37. Lee, S., Aronoff, S., *Science* (1967) **158,** 798.
38. Parish, R. W., *Enzymologia* (1968) **35,** 239.
39. Parish, R. W., *Z. Pflanzenphysiol.* (1969) **6,** 211.
40. Odhnoff, C., *Physiol. Plant.* (1957) **10,** 984.
41. Dutta, T. R., McIlrath, W. J., *Bot. Gaz.* (1964) **125,** 89.
42. McIlrath, W. J., Palser, B. F., *Bot. Gaz.* (1956) **118,** 43.
43. Odhnoff, C., *Physiol. Plant.* (1961) **14,** 187.
44. Jaweed, M. M., Scott, E. G., *Proc. W. Va. Acad. Sci.* (1967) **39,** 186.
45. Coke, L., Whittington, W. J., *J. Exp. Bot.* (1968) **19,** 295.
46. Saini, H. W., *et al., Curr. Sci.* (1969) **38,** 356.
47. Whittington, W. J., *J. Exp. Bot.* (1959) **10,** 93.
48. Shkol'nik, M. Y., *et al., Int. Congr. Biochem. V* (1961) **2,** 468.
49. Shkol'nik, M. Y., Solov'eve, E. A., *Bot. Zhur.* (1961) **46,** 161.
50. Shkol'nik, M. Y., Kasitsyn, A. V., *Dokl. Biol. Sci. Sect.* (1962) **144,** 622.
51. Shkol'nik, M. Y., *Tr. Bot. Inst., Akad. Nauk SSSR* (1970) **20,** 3.
52. Sherstnev, E. A., Kurilenok, G. V., *Dokl. Biol. Sci.* (1962) **142,** 50.
53. Sherstnev, E. A., Kurilenok, G. V., *Biol. Abstr.* (1965) **46,** 67808.
54. Borshchenko, G. P., Sherstnev, E. A., *Sov. Plant Physiol.* (1968) **15,** 602.
55. Sherstnev, E. A., *Tr. Bot. Inst., Akad. Nauk SSSR* (1970) **20,** 55.
56. Timashov, N. D., *Mikroelem. Sel. Khoz. Med.* (1967) **3,** 35.
57. Shiralipour, A., *et al., Crop Sci.* (1969) **9,** 455.
58. Borshchenko, G. P., *Tr. Bot. Inst., Akad. Nauk SSSR* (1970) **20,** 61.
59. Albert, L. S., *Plant Physiol.* (1965) **40,** 649.
60. Johnson, D. L., Albert, L. S., *Plant Physiol.* (1967) **42,** 1307.
61. Hundt, I. G., *et al., Albrecht-Thaer-Arch.* (1970) **24,** 725.

62. Rapota, V. V., *Fiz. Biokhim. Kul't, Rast.* (1970) **2**, 210.
63. Cory, S., *et al.*, *Phytochem.* (1966) **5**, 625.
64. Cory, S., Finch, L. R., *Phytochem.* (1967) **6**, 211.
65. Whittington, W. J., *J. Exp. Bot.* (1957) **8**, 353.
66. Slack, C. R., Whittington, W. J., *J. Exp. Bot.* (1964) **15**, 495.
67. Skok, J., *Plant Physiol.* (1957) **32**, 648.
68. Neales, T. F., *J. Exp. Bot.* (1959) **10**, 426.
69. Neales, T. F., *J. Exp. Bot.* (1964) **15**, 647.
70. Neales, T. F., Hinde, R. W., *Physiol. Plant.* (1962) **15**, 217.
71. Wildes, R. A., Neales, T. F., *J. Exp. Bot.* (1969) **20**, 591.
72. Albert, L. S., Wilson, C. M., *Plant Physiol.* (1961) **36**, 244.
73. Humphries, E. C., *Nature (London)* (1961) **190**, 701.
74. MacInnes, C. B., Albert, L. S., *Plant Physiol.* (1969) **44**, 965.
75. Yih, R. Y., Clark, H. E., *Plant Physiol.* (1965) **40**, 312.
76. Lee, S. G., Aronoff, S., *Plant Physiol.* (1966) **41**, 1570.
77. Wall, J. R., Andrus, C. F., *Amer. J. Bot.* (1962) **49**, 758.
78. Tehrani, G., *et al.*, *J. Amer. Soc. Hort. Sci.* (1961) **96**, 226.
79. Spurr, A. R., *Science* (1952) **116**, 421.
80. Perkins, H. J., Aronoff, S., *Arch. Biochem. Biophys.* (1956) **64**, 506.
81. Dear, J., Aronoff, S., *Plant Physiol.* (1965) **40**, 458.
82. Watanabe, R., *et al.*, *Arch. Biochem. Biophys.* (1961) **94**, 241.
83. Watanabe, R., *et al.*, *Phytochem.* (1964) **3**, 391.
84. Rajaratnam, J. A., *et al.*, *Science* (1971) **172**, 1142.
85. Krupnikova, T. A., *Tr. Bot. Inst. Akad. Nauk SSSR* (1970) **20**, 128.
86. Neales, T. F., *Nature (London)* (1959) **183**, 483.
87. Scholz, V. G., *Kulturpflanze* (1962) **10**, 63.
88. Kull, V., *Biochem. Physiol. Pflanz.* (1970) **161**, 330.
89. Augsten, H., Hundt, I., *Biol. Zentralbl.* (1970) **89**, 497.
90. Timashov, N. D., *Sov. Plant Physiol.* (1970) **17**, 838.
91. Spurr, A. R., *Science* (1957) **126**, 78.
92. Spurr, A. R., *Amer. J. Bot.* (1957) **44**, 637.
93. Wilson, C. M., *Plant Physiol.* (1961) **36**, 336.
94. Stanley, R. G., Loewus, F. A., in "Pollen Physiology and Fertilization," pp. 128, North-Holland Publ. Co., 1964.
95. Yih, R. Y., *et al.*, *Plant Physiol.* (1966) **41**, 815.
96. Mondy, N. I., *et al.*, *J. Food Sci.* (1965) **30**, 420.
97. Shkol'nik, M. Y., Kopman, I. V., *Tr. Bot. Inst. Akad. Nauk SSSR* (1970) **20**, 108.
98. O'Kelley, J. C., *Amer. J. Bot.* (1957) **44**, 239.
99. Vasil, I. K., in ref. 94, pp. 107.
100. Kumar, S., Hecht, A., *Biol. Plant.* (1970) **12**, 41.
101. Baker, J. E., *et al.*, *Plant Physiol.* (1956) **31**, 89.
102. Starck, J. R., *Acta Soc. Bot. Pol.* (1963) **32**, 619.
103. Lewin, J. C., *J. Exp. Bot.* (1966) **17**, 473.
104. Lewin, J., *J. Phycol.* (1966) **2**, 160.
105. Neales, T. F., *Aust. J. Biol. Sci.* (1967) **20**, 67.
106. Eyster, C., *Nature (London)* (1952) **170**, 755.
107. Bowen, J. E., *et al.*, *J. Phycol.* (1965) **1**, 151.
108. McIlrath, W. J., Skok, J., *Bot. Gaz.* (1957) **119**, 231.
109. Dear, J. M., Aronoff, S., *Plant Physiol.* (1968) **43**, 997.
110. Bowen, J. E., Gauch, H. G., *Plant Physiol.* (1966) **41**, 319.
111. Bowen, J. E., Gauch, H. G., *Amer. Fern J.* (1965) **55**, 67.

RECEIVED January 7, 1972.

8

Boron in Cultivated Soils and Irrigation Waters

F. T. BINGHAM

University of California, Riverside, Calif. 92502

A review paper covering forms of boron in soils, interactions between soil solution boron and adsorbed boron, adsorption–desorption processes, and relationships to plant nutrition is presented. Diagnostic criteria are given for chemical analysis of soils and irrigation waters in terms of boron status, i.e., *deficient, adequate, or excessive, and specified according to crop species group. The potential boron hazard of municipal sewage effluents to water supplies is discussed.*

Solution culture experiments conducted during the 1920's demonstrated that boron (B) is an essential nutrient element for higher green plants and that relatively low substrate concentrations of boron are phytotoxic to many plants. The range between beneficial and toxic concentrations is narrow. For example, solution culture concentrations of 0.05 to 0.10 μgram of B/ml are ordinarily safe and adequate for many plants whereas concentrations of 0.50–1.0 μgram of B/ml are frequently excessively high for boron sensitive plants. In addition, this research led to the identification and description of boron deficiency and excess visual symptoms and their relationship to plant tissue analysis (*1, 2*). Diagnostic criteria for soil and irrigation water analysis were subsequently taken into consideration, and although a number of problems were encountered owing to the complexities of soil systems, criteria were proposed for evaluating existing and potential boron management problems.

There are a number of limitations, however, which must be taken into account in evaluating the boron content of soils and waters in terms of food and fiber production. Boron reactions in soil–water systems, forms readily available to plants, and diagnostic criteria for soil and water analysis are examined in the following discussion.

Boron in Soils

Total Boron. Boron is present in organic matter, various soil minerals such as tourmaline, and in the soil solution in equilibrium with boron adsorbed on surfaces of soil particles. The total content of boron in soil includes all of the above-mentioned forms; however, the bulk of the boron comes from soil minerals. Hence, the boron content of soil is primarily related to the boron content of the parent material from which the soil was derived. Igneous rocks contain approximately 10 μgrams of B/gram whereas sedimentary rocks depending upon their genesis usually contain greater amounts. The relationship between rock source and boron content is brought out by the extensive survey of United States soils by Whetstone, Robinson, and Byers (*3*). They found an average content of 30 μgrams of B/gram for 200 soils with average values of 14 and 40 μgrams of B/gram for soils derived from igneous and sedimentary rocks, respectively. Soils from marine shales, in particular, were relatively rich in boron. Similar associations have been noted in soils of California. The boron content of soils weathered from alluvium eroded from marine shales in Kern County varies from 25 to 68 μgrams of B/gram whereas soils from granitic material contain 10 μgrams of B/gram or less (*4*). Additional information on boron contents of soils is presented by Swaine (*5*) and in review papers by Bradford (*6*), Mitchell (*7*), and Hodgson (*8*). The treatise by Philipson (*9*) on boron in plant and soil is recommended for additional details of boron reactions in soils and related nutritional characteristics.

Boron in Organic Matter. Although much of the boron in soils is associated with minerals resistant to weathering (*3*), boron is also contained in the organic fraction of soils. Little is known, however, of the reactions and availability of boron in soil organic matter other than that the quantity is small and is restricted to the surface horizon of soils primarily. As this organic matter fraction mineralizes, the boron redistributes in the soil–water system, becoming available in part for plants (*10*).

Adsorbed Boron. Boron precipitated and adsorbed on surfaces of soil particles is probably of greater importance to plant growth because of equilibria existing between adsorbed and soluble boron. A substantial proportion of the boron added to soil either as a component of fertilizers or in irrigation water is adsorbed by certain soil materials, the balance remaining in the soil solution. This soil solution concentration is especially important to plant nutrition because of its immediate availability to plants. Plants respond primarily to the soil solution boron, independently of the amount of boron adsorbed by soil (*11*). Consequently, conditions affecting equilibria between adsorbed and soluble boron are highly relevant to considerations of plant nutrition and diagnostic procedures for soils and irrigation waters.

Briefly, boron adsorption by soils depends upon the boron concentration (especially that of the borate ion) of the solution in equilibrium with the solid phase of soil, pH of the soil system, and number of active adsorption sites per unit weight of soil (*12–18*). These adsorption sites are associated with broken Si–O and Al–O bonds exposed at edges of aluminosilicate minerals (*19, 20*) and also with surfaces of amorphous hydroxide materials present in weathered soils such as allophane, and hydroxyaluminum and iron compounds (*12–18*). The adsorption sites in arid zone soils, according to Rhoades *et al.* (*21*), are associated with magnesium hydroxide clusters and coatings that form on the exposed surfaces of ferromagnesium minerals and micaceous layer silicates. Although layer silicate clays exhibit an affinity for boron, the greater part of the adsorption is ascribed to sesquioxide coatings on surfaces of the clay rather than to the exposed Si–O and Al–O bonds (*13–15*).

Boron adsorption occurs independently of variations in pH within the acid range. Increases in pH within the alkaline range result in increased adsorption, with maximum adsorption taking place at about pH 9.0 (*13–16, 18, 22*). The above pH effects on boron suggest that it is adsorbed as molecular boric acid under acid conditions, and borate ion is adsorbed as the pH approaches 9.0 (*13–16, 18, 22*). A recent report from Australia (*23*) proposes that the above adsorption maximum for boron at pH 9 is ascribed to protons dissociating from boric acid, subsequently reacting with surface hydroxyl groups of oxide surfaces (*viz.*, hydroxy aluminum compounds) to form positively charged sites for adsorbing borate ions. Thus maximum adsorption takes place at the pK_a of boric acid. Monosilicic has a comparable pK_a, and maximum adsorption occurs in the same pH range (*24*) as for boric acid.

Boron adsorption takes place independently of concomitant adsorption of other anions. For example, adsorption studies with allophanic soils from central Mexico (*22*) and southern Chile (*18*) revealed no effect from the simultaneous adsorption of sulfate or phosphate on boron adsorption.

Regarding the quantity of boron adsorbed, adsorption maxima calculated with the Langmuir adsorption equation (*16, 17, 25–27*) vary from approximately 10 up to 100 μgrams of B/gram of soil. Soils derived from volcanic ash deposits adsorb unusually large amounts of boron (*16, 18*) because of their enriched contents of amorphous materials with affinity for anions.

Under ordinary conditions, soils in the field contain small amounts of soluble and adsorbed boron, and thus it is difficult to obtain a reliable precise measurement of adsorbed boron. Investigators frequently use a

Langmuir adsorption isotherm technique with soil samples (*25–27*), but this may have its failings too (*28*).

An indication of the quantity adsorbed may be obtained from observations of elution or leaching characteristics. With such a technique. alluvial soils in western Kern County, Calif., were found to contain as much as 30% of the total boron in the adsorbed or easily leached fraction (*4*). However, these Kern County soils are unusually high in boron, and hence their boron status is not representative of arable soils in general. Subsequent reclamation experiments with salt-affected, high boron soils demonstrated the feasibility of leaching excessive boron by sprinkler irrigation (*29*) and by basin irrigation (*30*). Although boron leaches, it does not ordinarily leach out of the profile as readily as chloride, nitrate, and sulfate salts (*29*). Boron reclamation was essentially a matter of the depth of water leached through the soil. Fallowing these reclaimed soils for 12 months did not result in an increase in soluble boron. However, laboratory leaching experiments by Rhoades *et al.* (*28*) with high boron soils showed that boron was leached or extracted relatively easily, but it tended to solubilize to toxic levels once the leaching or extraction ceased. These investigators were able to estimate the extent of boron removal needed to "reclaim" soil through a simple laboratory extraction using either a mannitol–$CaCl_2$ solution or a resin.

Although plants absorb boron directly from the soil solution, the adsorbed boron provides a source of boron to maintain the soil solution level. Therefore, soil tests for available boron (that available to plants) are designed to measure the boron in the soil solution as well as a substantial part of the adsorbed or readily soluble boron.

Water-Soluble Boron. Under field conditions, water-soluble boron is the boron present in the soil solution which may be visualized for this discussion as the solution bathing the roots. This boron fraction is available to plants for uptake and assimilation. The technique for obtaining this solution consists of bringing the soil to a given moisture level, allowing a period for equilibration—*i.e.*, 24 hr, and displacing the solution. The boron concentration of this solution is generally low, less than 0.1 or 0.2 μgram of B/ml; however, that of arid zone soils may be considerably higher.

Boron deficiency is widespread with a variety of crops, according to Berger (*31*). Deficiencies have been identified in at least 41 states of the United States, mainly in the humid zones. Diagnostic procedures include soil and plant analysis as well as recognition of specific symptoms characteristic of plants under boron stress. The hot water extraction procedure for available boron by Berger and Truog (*32*) is one of the accepted procedures for estimating available boron. Briefly, it consists of boiling a 1:2 soil to water suspension for 5 min in a reflux condenser and separating the solution phase to determine boron. The quantity of extracted boron is a measure of available boron for a given group of crop species. For example, crop species such as alfalfa, sugar beet, and celery are classified as high boron crops and thus call for a higher soil analysis value than the low boron crops such as grains, grasses, and soybeans (*10*). Table I contains a summary of soil analysis by the Berger–Truog method (*32*) for appraising the boron fertilizer needs of field and vegetable crops. In general, concentrations of 0.5 μgram of B/gram or lower are indicative of a boron deficiency, depending upon the crop species.

Table I. Diagnostic Criteria for Boron Nutrition of Field and Vegetable Crops Based upon Hot Water Extraction of Soil Boron (*10*)

Boron Content of Soils for Optimum Growth		
0.1 μgram of B/gram	*0.1–0.5 μgram of B/gram*	*0.5 μgram of B/gram*
Small grain	Tobacco	Apple
Corn	Tomato	Alfalfa
Soybean	Lettuce	Clovers
Pea and bean	Peach	Beets
Strawberry	Pear	Turnips
Potato	Cherry	Cruciferae
Grass	Olive	Asparagus
Flax	Pecan	Radish
	Cotton	Celery
	Sweet potato	Rutabaga
	Peanut	
	Carrot	
	Onion	

Soil analysis for appraisal of boron toxicity in soils also entails measurement of water-soluble boron but by a different extraction procedure—the saturation-extract procedure (*33*). This technique was proposed by the staff of the U. S. Salinity Laboratory for diagnosing saline and alkali soils, including high boron soils. The technique is simple: a soil is saturated with water, and after a 24-hr equilibration period, the solution is extracted under vacuum. The boron concentration of a saturation extract is comparable with that of the soil solution. According to Eaton's sand culture experiments with a range of crop species (*1, 2*), sensitive plants (citrus, avocados, various deciduous fruit trees, etc.) are injured from boron with substrate concentrations as low as 1 μgram of B/ml. He noted that many field and vegetable plants were able to grow normally with concentrations above 10 μgrams of B/ml. Thus, there is a relatively broad range of boron concentrations that plants tolerate depending upon the plant species. Although observations of crop species under field conditions with variable soil solution or saturation-extract boron concentra-

Table II. Toxic Boron Concentrations of Saturation Extracts for Sensitive, Semitolerant, and Tolerant Crop Species[a]

Saturation-Extract Boron, μgrams of B/ml		
0.5–1.0	*1.0–5.0*	*5.0–10.0*
Sensitive	*Semitolerant*	*Tolerant*
Citrus	Lima bean	Carrot
Avocado	Sweet potato	Lettuce
Apricot	Bell pepper	Cabbage
Peach	Oat	Turnip
Cherry	Milo	Onion
Persimmon	Corn	Broad bean
Fig	Wheat	Alfalfa
Grape	Barley	Garden beet
Apple	Olive	Mangel
Pear	Field pea	Sugar beet
Plum	Radish	Palm
Navy bean	Tomato	Asparagus
Jerusalem artichoke	Cotton	
Walnut	Potato	

[a] Listed in each category according to susceptibility to boron injury (*viz.*, citrus is more sensitive than walnut, lima bean more than potato, etc.).

tions are limited, plant response to soluble boron parallels that observed by Eaton with sand cultures. Hence, diagnostic criteria in use are essentially those deduced from his studies. Table II contains crop species listed into sensitive-, semitolerant-, and tolerant-crop groupings for saturation-extract concentrations respectively of 0.5–1.0, 1.0–5.0, and 5.0–10.0 μgrams of B/ml. Admittedly, variations in soil properties such as pH, organic matter content, and clay content may alter to a degree the availability of a given level of soluble boron (*11, 34–36*), but at this time there is not sufficient information available to compensate for such variations. The question of boron excess can be further resolved by examining suspect plants for specific injury symptoms and elevated plant tissue concentrations of boron (*1, 2*).

Boron in Irrigation Waters

The first observation of boron toxicity caused by irrigation water was made in 1925 by Kelley and Brown (*37*). They associated specific injury symptoms with abnormally high concentrations of boron in leaves of walnut and citrus orchards in southern California. They noted that orchards irrigated with water containing 1.0 μgram of B/ml or more frequently exhibited injury symptoms. In 1931, Scofield and Wilcox (*38*) concluded from additional studies of water supplies in southern California that the critical (injurious) concentration of boron in irrigation waters was 0.5–1.0 μgram of B/ml for sensitive crops. Later, in 1935, Eaton (*2*) extended the information on appraisal of boron content of irrigation waters from a study of irrigation waters of the Central Valley, Calif. Eaton proposed critical concentrations of 0.5–1.0 and 1.0–2.0 μgrams of B/ml, respectively, for sensitive and semitolerant crops. The above observations, dating back to 1925, constitute the basis for water quality criteria.

Perhaps the most widely accepted irrigation water criteria are those published in 1954 by the staff of the U. S. Salinity Laboratory (*33*). Criteria for boron include possible limits for four classes of irrigation waters according to sensitive-, semitolerant-, and tolerant-crop species. More recently, Wilcox (*39*) published a modified set of critical concentrations which are somewhat more flexible in that fewer categories are involved. He proposed the following for irrigation waters:

Crop Species	*Critical Boron Concentration, μgram of B/ml*
Sensitive	0.3–1.0
Semitolerant	1.0–2.0
Tolerant	2.0–4.0

These groupings of crop species are identical with those listed in Table II.

With reference to citrus, Chapman (*40*) has suggested concentrations of 0.5, 1.0, and 2.0 μgrams of B/ml for possible-, definite-, and serious-hazard classes for irrigation waters.

A serious limitation of such boron criteria is the lack of provision for difference in soils, irrigation management, and climate. Soils differ in their capacity for adsorbing boron, and irrigation and rainfall characteristics likewise exert an effect on the distribution of available boron in soils. Additional details are presented in discussions of criteria for irrigation water quality by Bernstein (*41*) and Rhoades (*42*).

Fortunately, the majority of surface water supplies have boron concentrations ranging from 0.1 to 0.3 μgram of B/ml (*33*) and hence are ordinarily safe to use for irrigation with respect to boron. Well waters, however, are more variable in the aggregate, and excessive amounts of boron are often encountered. Underground water supplies constitute approximately 50% of the water used for irrigation in the western United States.

Another source of excessive boron is sewage effluent. Boron originating from the householder's use of certain laundry products as well as from some industrial plants is becoming a matter of concern to downstream users of surface waters degraded by discharge of processed sewage effluent. An example is the Santa Ana River basin in southern California. This area is being urbanized rapidly and, in general, at the expense of irrigated agriculture. The return flow from irrigated lands entering the Santa Ana River system is decreasing volumewise while that from municipalities (sewage effluent) is increasing. The waste discharge of communities in this area contains 0.75–1.50 μgrams of B/ml, depending upon the city and season. Although there are a variety of sources of this boron, the amount from households accounts for a substantial portion of the total, perhaps as much as 50% (*43*). With increasing population, this contribution is expected to become correspondingly greater as long as laundry products containing boron are used in the present proportion to population. Consequently, the local Regional Water Quality Control Board considers an incremental addition (increase in sewage effluent above that of incoming fresh water) of 0.5 μgram of B/ml excessive from the point of view that such waters, upon being discharged back into the Santa Ana River system, degrade the quality of the river supply for irrigation. Currently the incremental additions to the Santa Ana River system average about 0.8 μgram of B/ml (*44*). Water quality criteria for return flow and sewage effluent supplies should also be based upon considerations of rainfall diluting the boron concentration of the degraded water supply (downstream from discharge), and the hazard of boron in sewage effluent is greatly diminished or actually eliminated under conditions where the effluent percolates through soil.

Literature Cited

1. Eaton, F. M., *J. Agr. Res.* (1944) **69,** 237.
2. Eaton, F. M., *U.S.D.A. Tech. Bull.* (1935) **448.**
3. Whetstone, R. R., Robinson, W. O., Byers, H. G., *U.S.D.A. Tech. Bull.* (1942) **797.**
4. Bingham, F. T., Arkley, R. J., Coleman, N. T., Bradford, G. R., *Hilgardia* (1970) **40,** 193.
5. Swaine, D. J., "Commonwealth Bureau of Soil Science Technical Communication," No. 48, Herald Printing Works, York, England, 1955.
6. Bradford, G. R., "Diagnostic Criteria for Plants and Soils," Chapter 4, pp. 33–61, University of California Division of Agriculture Sciences, 1966.
7. Mitchell, R. L., "Chemistry of Soil," Chapter 8, pp. 320–368, ACS Monograph No. 126, 2nd ed, Reinhold, New York, N. Y., 1965.
8. Hodgson, J. F., *Advan. Agronomy* (1963) **15,** 119.
9. Philipson, T., *Acta Agr. Scand.* (1953) **3** (2), 121.
10. Berger, K. C., *Advan. Agronomy* (1949) **1,** 321.
11. Hatcher, J. T., Blair, G. Y., Bower, C. A., *Soil Sci.* (1959) **88,** 98.
12. Hatcher, J. T., Bower, C. A., Clark, M., *Soil Sci.* (1967) **104,** 422.
13. Sims, J. R., Bingham, F. T., *Soil Sci. Soc. Amer. Proc.* (1967) **31,** 728.
14. Sims, J. R., Bingham, F. T., *Soil Sci. Soc. Amer. Proc.* (1968) **32,** 364.
15. Sims, J. R., Bingham, F. T., *Soil Sci. Soc. Amer. Proc.* (1968) **32,** 369.
16. Bingham, F. T., Page, A. L., Coleman, N. T., Flach, K., *Soil Sci. Soc. Amer. Proc.* (1971) **35,** 546.
17. Okazaki, E., Chao, T. T., *Soil Sci.* (1968) **105,** 255.

18. Schalscha, E. B., Bingham, F. T., Galindo, G. G., Galvan, H. P., *Soil Sci.* (1973) **116,** in press.
19. Fleet, M. E. L., *Clay Minerals* (1965) **6,** 3.
20. Couch, E. L., Grim, R. E., *Clays, Clay Minerals* (1968) **16,** 249.
21. Rhoades, J. D., Ingvalson, R. D., Hatcher, J. T., *Soil Sci. Soc. Amer. Proc.* (1970) **34,** 938.
22. Bingham, F. T., Page, A. L., *Soil Sci. Soc. Amer. Proc.* (1971) **35,** 892.
23. Hingston, F. J., Poshner, A. M., Quirk, J. P., *Search* (1970) **1,** 324.
24. McPhail, M., Page, A. L., Bingham, F. T., *Soil Sci. Soc. Amer. Proc.* (1972) **36,** 510.
25. Hatcher, J. T., Bower, C. A., *Soil Sci.* (1958) **94,** 55.
26. Biggar, J. W., Fireman, M., *Soil Sci. Soc. Amer. Proc.* (1960) **24,** 115.
27. Singh, S. S., *Soil Sci.* (1964) **98,** 383.
28. Rhoades, J. M., Ingvalson, R. D., Hatcher, J. T., *Soil Sci. Soc. Amer. Proc.* (1970) **34,** 871.
29. Bingham, F. T., Marsh, A. W., Branson, R., Mahler, R., Ferry, G., *Hilgardia* (1972) **41,** 195.
30. Reeve, R. C., Pillsbury, A. F., Wilcox, L. V., *Hilgardia* (1955) **24,** 69.
31. Berger, K. C., *J. Agr. Food Chem.* (1962) **10,** 178.
32. Berger, K. C., Truog, E., *Ind. Eng. Chem., Anal. Ed.* (1939) **11,** 540.
33. Staff, U. S. Salinity Laboratory, U.S.D.A. Handbook (1954) **60.**
34. Bingham, F. T., Elseewi, A., Oertli, J. J., *Soil Sci. Soc. Amer. Proc.* (1970) **34,** 613.
35. Stinson, C. H., *Soil Sci.* (1953) **75,** 31.
36. Wear, J. I., Patterson, R. M., *Soil Sci. Soc. Amer. Proc.* (1962) **26,** 344.
37. Kelley, W. P., Brown, S. M., *Hilgardia* (1928) **3,** 445.
38. Scofield, C. S., Wilcox, L. V., *U.S.D.A. Tech. Bull.* (1931) **264.**
39. Wilcox, L. V., U.S.D.A. Information Bull. (1960) **211.**
40. Chapman, H. D., "Citrus Industry," Vol. 2, pp. 127–289, University of California Division of Agriculture Sciences, 1968.
41. Bernstein, L., *Amer. Soc. Testing Mater. Spec. Tech. Publ.* (1967) **416.**
42. Rhoades, J. D., *Soil Sci.* (1972) **113,** 277.
43. "Special Report on Boron," Santa Ana Regional Water Quality Control Board, Riverside, Calif., 1965.
44. "Final Report on Task IV-3," Santa Ana Watershed Planning Agency, Riverside, Calif., 1971.

RECEIVED January 19, 1972.

9

A Macroreticular Boron-Specific Ion-Exchange Resin

ROBERT KUNIN

Rohm and Haas Co., Independence Mall, West, Philadelphia, Pa. 19105

Recent laboratory studies and field operations have demonstrated the excellent performance of a macroreticular boron-specific ion-exchange resin, Amberlite XE-243, for deborating several process streams, sea water, and irrigation water. One large plant is now operating successfully on sea water. Further studies on this product are currently in progress in several areas throughout the world.

Although problems associated with boron (B) in water supplies have been recognized for years in many sectors of the agricultural community, until recently these problems have been limited primarily to arid areas. With the widespread use of boric acid and its salts in domestic and industrial detergent and cleansing formulations, the presence of boron in our waters has taken on the aspects of a more general problem.

Boron in the soil and water exists either as boric acid or as the borate ion. Although the levels generally present pose no problems to animal life, in many areas, present levels of boron in water supplies can be deleterious to agricultural crops. Boron is a typical case in which there exists a small difference between a deficiency level and a toxicity level in the water used for agricultural purposes. Boron phytotoxicity can occur with levels above 2 ppm, and deficiency can occur for many plant species if the levels are below 0.5–1.0 ppm of B.

Boron is also a problem in some areas of industry. Magnesium oxide (MgO) used as a refractory material or magnesium chloride ($MgCl_2$) used to produce magnesium metal must be low in boron. Because of the high atomic cross section of boron, the boron content of many materials of construction and chemicals used in atomic energy must also be extremely low.

Because of the current widespread presence of boron in most waterways and the need to remove it from many irrigation waters and industrial chemical streams, the author and his associates (*1*) embarked upon a program to develop a boron-specific ion-exchange resin. Although boron as boric acid can be removed from water with strong base anion-exchange resins during deionization, all other ionic species are also removed, rendering the operation uneconomical if boron is the only objectionable constituent. The first result of the program geared to the development of a boron-selective ion-exchange resin was Amberlite XE-143, a crosslinked microreticular or gel-type copolymer based upon the amination of chloromethylated styrene–divinylbenzene with *N*-methylglucamine,

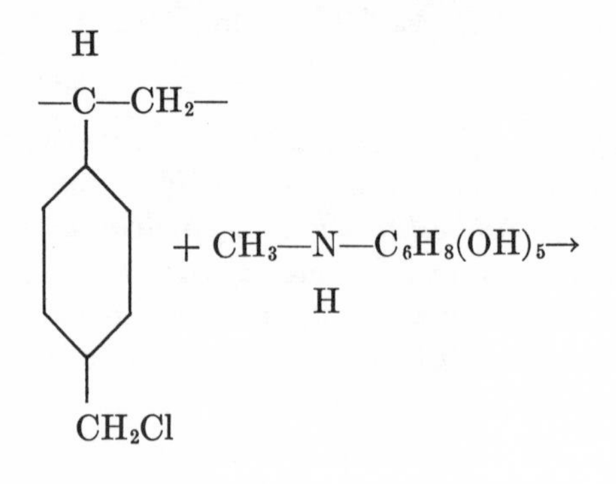

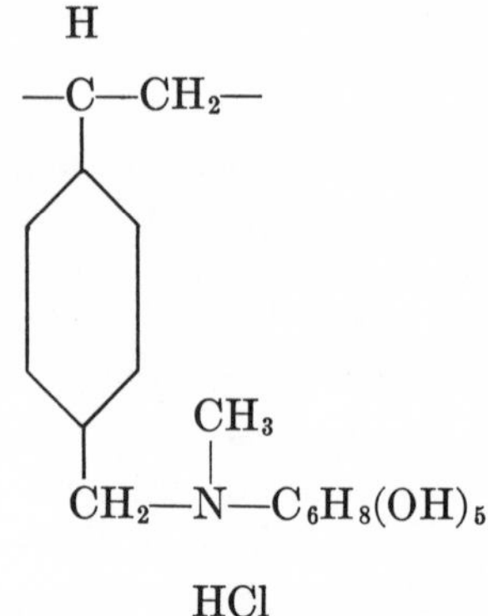

Extensive studies in the laboratory and in the field verified the excellent boron selectivity of this ion-exchange resin; however, some serious deficiencies in kinetics and physical stability were encountered. More recently an analog of this product, Amberlite XE-243 (Amberlite XE-243 is now commercially available under the designation Amberlite IRA-943), based upon a macroreticular styrene–divinylbenzene, was developed which overcame these deficiencies. The difference between the two structures is only physical. The macroreticular structure of the new boron-selective ion exchanger is similar to that of other macroreticular ion-exchange resins that have been previously described (*2, 3*).

In contrast to the microreticular or gel-type ion-exchange resins, the macroreticular ion-exchange resins possess a pore structure superimposed upon the crosslinked copolymer structure that is similar to that of the classical adsorbents such as the aluminas, silicas, and the carbons. The macroreticular structure serves a host of purposes in ion-exchange technology. It permits one to prepare ion-exchange resins of unusual physical and chemical stabilities and to introduce functional groups that ordinarily could not have been introduced effectively into a microreticular crosslinked copolymer structure.

The difference in physical structure between the Amberlite XE-243 and the older Amberlite XE-143 is readily apparent from the electromicroscopic photomicrographs described in Figures 1 and 2. The general characteristics of Amberlite XE-243 are described in Table I.

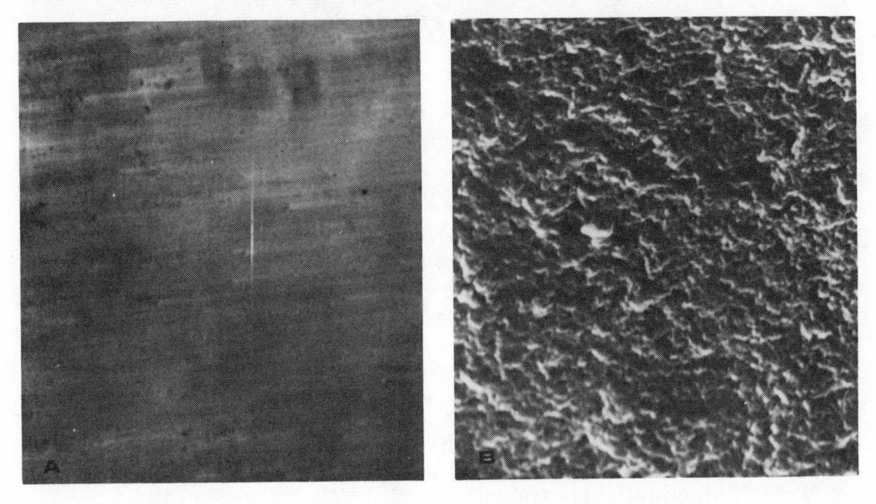

Figure 1. Electron micrograph of (A) conventional gel and (B) macroreticular structures

The columnar operating capacities of Amberlite XE-243 for removing boron from a synthetic irrigation water (1000 ppm NaCl and 10 ppm B) are described in Figure 3. The column studies were conducted in 50-ml burets containing 20 ml (backwashed and drained) of resin. A flow rate of 1–4 gallons per cubic foot per minute was used for exhaustion and 1 gallon per cubic foot per minute for regeneration. The influent pH was 6.9. An endpoint of 1 ppm of B was arbitrarily chosen. Throughout the study, boron was determined by the carminic acid procedure. Included in Figure 3 are data more recently obtained by Aerojet General Corp. (*4*), confirming the accuracy and reliability of the data as presented. The capacity of Amberlite XE-243 for boron is independent of pH and ionic strength, and a recent report (*5*) has now confirmed the unique performance of Amberlite XE-243 in removing the objectionable traces of boron in sea water being used for magnesium oxide production and mag-

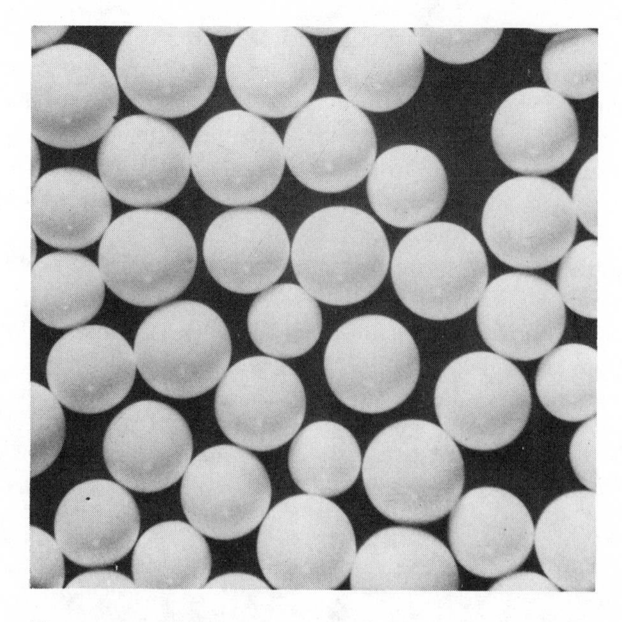

Figure 2. Photomicrograph of beads of Amberlite XE-243

Table I. General Characteristics of Amberlite XE-243

Moisture content, %	58
Apparent density, pounds-/ft.^{3}	43
Anion-exchange capacity, meq/ml.	2.7
Macroreticular porosity, %	48

nesium chloride brines being processed for magnesium metal. Unpublished reports have noted the unique performance of the same product for removing traces of boron from sewage treatment plant effluents. For sewage plant effluents, Amberlite XE-243 in addition to removing boron, reversibly removes some COD values that elude the carbon and coagulation stages.

With respect to further developments concerning the use of Amberlite XE-243 for removing boron from irrigation waters and industrial liquors, much of the activity has involved the regeneration step. Previous studies (*1*) suggested a two-stage regeneration consisting of H_2SO_4 (3 pounds/ft^3), water rinse, NaOH (4 pounds/ft^3), and a final water rinse. Although this regeneration procedure has proved to be excellent in all cases, the use of the two reagents in the amounts required has been uneconomical in some instances. Tests performed by Aerojet General Corp. (*4*) have demonstrated that the H_2SO_4 acid dosage could be

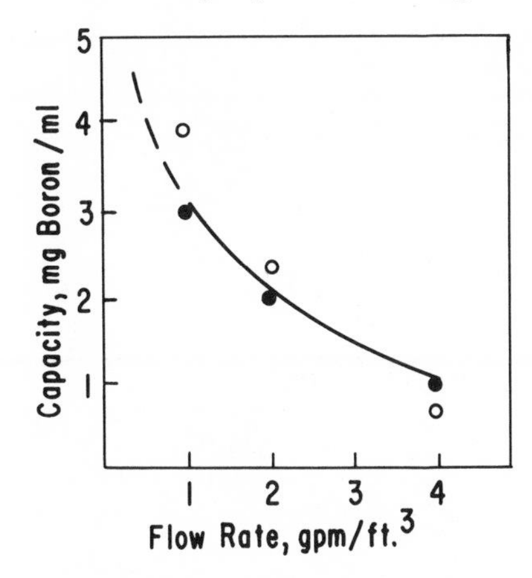

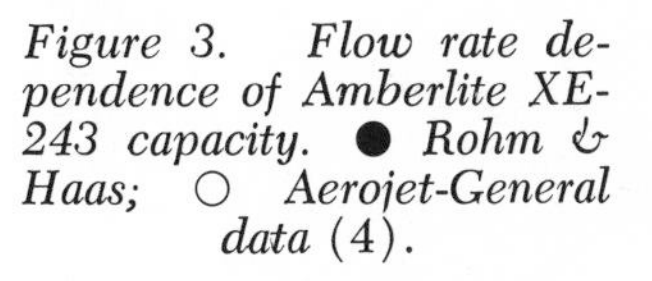
Figure 3. Flow rate dependence of Amberlite XE-243 capacity. ● *Rohm & Haas;* ○ *Aerojet-General data* (4).

reduced to 1.9 pounds of H_2SO_4 and the caustic eliminated with but a 14% lowering in boron capacity. In addition, HCl can be substituted for the H_2SO_4. Further, the waste acid could be recycled several times after permitting boric acid to crystallize from the concentrated regenerant "cuts."

With the increased physical stability of the macroreticular Amberlite XE-243 and the economic advances achieved with the regeneration step, it now appears that the average irrigation water and tertiary sewage effluent can be deborated at overall costs well below $0.03 per 1000 gallons.

Further studies with Amberlite XE-243 for the aforementioned applications are currently in progress in several areas throughout the world.

Literature Cited

1. Kunin, R., Preuss, A. F., *Ind. Eng. Chem., Prod. Res. Develop.* (1964) **3,** 304.
2. Kun, K. A., Kunin, R., *J. Polymer Sci.* (1964) **B2,** 389, 587.
3. Kun, K. A., Kunin, R., *J. Polymer Sci.* (1967) **C16,** 1457.
4. Aerojet-General Corp., OSW Rept. (Jan. 1970) **1271-F.**
5. *Chem. Week* (Aug. 16, 1972), 37.

RECEIVED January 7, 1972.

INDEX